THE OXFORD PAPERS

Proceedings of the British Association of Paper Historians Fourth Annual Conference
Held at St. Edmund Hall, Oxford, 17 - 19 September 1993

Edited by Peter Bower

British Association of Paper Historians
1996

The British Association of Paper Historians
Individuals are invited to join the Association at £16 pa. Organisations and Institutions can also become members at a rate of £35 pa. UK students in full-time education can join for £10 pa. The rates for members outside the UK are £20 for individual members and £40 for organisations and institutions.

For further information contact:
Frances Wakeman, BAPH Membership Secretary, 2 Manor Way, Kidlington, Oxfordshire OX5 2BD.

Published by The British Association of Paper Historians, 64 Nutbrook Street, London SE15 4LE, 0171 732 0125

British Library Cataloguing in Publication Data
A Catalogue record for this book is available from the British Library.

ISBN 095 25757 0 1

Typeset and Printed by Upstream, 1 Warwick Court, Choumert Road, London SE15, 0171 358 1344
Bound by J. Muir, Binders, Blackheath
Text printed on Printmaster Plus 90gsm, and cover on Avalon White Wove 250gsm, provided by Inveresk Fine Papers, Caldwells Mill.

Cover Illustration:
Details from *The Paper Mill at Wolvercote, Oxfordshire, belonging to James Swann, Esquire,* watercolour by James Buckler, 1826. (Oxford University Press)

Dedicated to the memory of the late Robert Hopkins who taught me how to see paper. **PB**

CONTENTS

The various advertisments found in the pages of this
book are taken from issues of the *Paper Trade Review*,
published in *The British and Colonial Printer*, between
January and September 1882. (St Bride Printing Library)

ACKNOWLEDGEMENTS

A book such as this owes much to many people, many of whose specific help is acknowledged by the individual authors. As editor, my greatest thanks go to my fellow-authors: this book is the product of their commitment to a fascinating subject and without them all it would not have been possible. Particular thanks must also go to Sally Bower, Mollie Chilton, Alan Crocker, Phil Crockett, Harry Dagnall, Richard Hills, George Mandl, James Mosley and all at Upstream for making this volume possible.

We are especially grateful to Stefan Kay, Group Managing Director of Inveresk, Tom Sneddon, Commercial Director of Inveresk Fine Papers Division, Marlene Scott and Laura Henderson of Inveresk's Caldwells Paper Mill, Fife, for their very generous provision of the text and cover papers for this volume. Their active and generous sponsorship is not restricted to this volume. They have provided sufficient paper for the following two volumes in the *Studies in British Paper History* series, *The Exeter Papers* and *The London Papers*, which are currently in preparation, as well as providing sufficient paper for several future issues of *The Quarterly*.

INTRODUCTION

Several of the papers presented at earlier BAPH conferences have appeared as articles in *The Quarterly*, but the content and quality of the papers being given and our growing membership has led to the Association deciding to publish the Conference Proceedings as separate books. This volume, *The Oxford Papers*, the collected papers of the Conference held at St Edmund Hall, Oxford in September 1993, is the first of a projected series, entitled *Studies in British Paper History*. Volumes II and III, the proceedings of the 1994 Exeter and the 1995 London Conferences, respectively, are both currently in preparation. It is envisaged that this series will not be restricted to BAPH Conference papers but will also publish book length research into various aspects of British Paper History. There are currently several texts under consideration.

The long and very distinguished history of the paper industry is increasingly becoming an area of intense research throughout the world. More and more people working in many disciplines: art history, paper conservation, social, economic, industrial and scientific history and forensic science, are finding that they need to understand and describe particular papers and their characteristics, and understand the processes, technologies and people involved. Of equal importance is an understanding of the social and economic pressures and influences that have led to these specific developments in the industry and its products. Like paper itself, which is nothing unless it is being used, Paper Historical Research has an importance way beyond its own field. Increasing use is being made of the results of such research by people working in other disciplines, particularly for the dating and attribution of works of art, documents and other paper artefacts. A knowledge of paper history is becoming essential to a proper understanding of the precise materials and technologies used; as well as contributing greatly to specific areas of economic, literary, industrial, technological and social history.

The Oxford Conference began with a visit to Wolvercote Mill where David Bossom, the technical manager, and his staff made us very welcome. Wolvercote Mill began operation in 1674 and for a long period of its history was owned by the Oxford University Press. We then moved on to the Bodleian Library where Edward Simpson and Pascale Regnault showed us various items from their collections and there was much discussion of the various problems facing paper conservators as a result of earlier papermaking practices. A brief visit to Duke Godfrey's Library was followed by Julie Anne Lambert showing us some of the unsurpassed collection of printed ephemera, particularly those items relating to paper and its history, in the *Johnson Collection*. Johnson was printer to the Oxford University Press from 1925-46 and his collection is now housed in the Bodleian.

The conference had two themes, the paper history of Oxfordshire and the little-researched use of straw in papermaking. The first session consisted of two papers giving an introduction to the use of straw as a raw material for papermaking in the 19th century. Richard Hills began with Matthias Koops, the first person to try straw pulp as a commercial proposition in 1800. His construction of a large steam-driven mill on the banks of the Thames, at Millbank, ended in catastrophic failure. The expansion in the demand for paper in the first half of the nineteenth century had led to an increase in the price of rags and consequently to renewed interest in alternative fibre sources such as straw. Hills examined how this interest was reflected in the applications for British Patents during this period. Around 1850 there was a sudden rise in patent applications relating to the preparation of cellulose from straw, principally by chemical means. One successful papermaker in this field was Charles Townsend Hook, who used straw extensively to produce fine quality papers. Hills concluded with a brief look at the use of straw during the First World War, when other fibre sources became scarce.

In the second paper of this session, Peter Bower continued the story of straw usage in the nineteenth century by discussing the part it has played in European papermaking since its earliest use. This earliest usage involved various cereal straws being used as fillers, in combination with either low-grade linen rags, or with blends of low-grade, and often coloured, linen rag and old hemp ropes, in the production of brown, brown-white or some blue wrappings. This was a very different procedure to the way straw was to be used in the nineteenth and twentieth centuries, where the natural 'wetness' of straw fibre during beating has been exploited to produce low bulk and a certain harshness, or crispness, in the sheet, characteristics that are considered particularly useful in 'bonds', 'banks' and 'loans', where a crisp handle and firmness in the sheet are of great importance. The nineteenth century saw much discussion on a wide variety of alternative fibres. Bower concentrated on using people's own words to describe how they saw the possibilities and pitfalls of straw as a papermaking raw material in the nineteenth century, including the arguments about the actual, as opposed to theoretical, availability of large amounts of straw for papermaking.

Frances Wakeman began the next session by discussing Papermaking in the Oxford Area, in particular the histories of Hampton Gay, Eynsham and Sandford Mills and discussed the influence of the Oxford University Press on papermaking in the Oxford area. A visit to the new Oxford University Press Museum with its splendid displays of type, copperplates, artefacts and ephemera, was followed by Peter Foden, archivist of the OUP, unravelling the fact and fiction behind the 'invention' and production of *Oxford India Paper*. A highlight of this visit was the special opening of a sealed notebook containing notes on the development of the furnish used for *Oxford India Paper* and various experiments for its manufacture. Part of the pulping process, which all in all took some 48 hours, involved boiling 30 cwt of tarred hemp hawser coil, cut into 4 inch lengths, in 4 cwt of lime for 10 hours.

The afternoon session began with James Brander's exploration of Straw as a papermaking fibre. He reviewed the historical background, to show that, historically, it has only really ever been used as a substitute pulp. He then summarised its modern utilisation, showing how straw has, in recent times, generally been used in countries without full access to the world's commodity markets. Brander then reviewed the plant structure and its constituents as well as the physical and chemical characteristics of various cereal straws, particularly when compared with woods. He itemised various papermaking characteristics to show how straw can give some valuable papermaking properties. This wide-ranging paper then covered the difficulties involved in both the production and utilisation of straw pulps, and the recovery problems faced, especially with reference to small mills. Brander used the example of a small mill producing corrugated medium to discuss questions of capital and operating costs. Current research activities in this area were summarised, and finally, some comments were made on the likely development of straw as a pulp material.

This was followed by Phil Crockett's paper on the development of the hollander beater. He outlined various beating methods, including stampers, kollergangs, hollanders and refiners, explaining the working mechanisms and some of the problems that have been encountered. The end of the nineteenth century and the early years of the twentieth century saw many fascinating developments in the evolution of the design of the basic hollander, some successful, some less so, as people tried to improve the quality, quantity and rate of beating of this machine that has contributed so much to the development of the industry over so long a period.

After the annual dinner, Members' contributions included Penny Jenkins' presentation which illustrated some of the technical problems faced by paper conservators, resulting from the use of straw in the production of board and paper. These problems mainly relate to the poor quality cheap boards used by picture framers and mounters to lay down works of art. The degradation of unbleached, acidic, cereal strawboards can directly affect the visual and chemical stability of any art work laid down onto such boards. Jenkins then showed examples of the methods used for the removal of such backing boards. The same degradation can also be seen in some cheaper papers, containing bleached or processed straw, commonly used in the nineteenth century and in periods of war in this century. These papers can darken and may become brittle. However, many papers still look and handle well and respond agreeably to simple conservation wet treatments. Jenkins considers that there are many positive aspects about straw and that it is easy to see why artists sometimes chose straw papers for their warm buff tonality and interesting tonal qualities.

Russell Jones then spoke briefly and humorously of the range of publications one needed to search to keep track of the detail of Paper History and its study. Robin Clarke then talked further on Oxfordshire Papermaking, in particular concentrating on Shiplake Mill and on Ponsonby's design for a waterwheel that in some details was the beginnings of a design for a turbine, (Robin Clarke's contribution was unfortunately not available for publication in this volume). The evening session continued with Barry Watson's showing of 25 photographs taken by William Raitt of Kashmiri papermaking in the Arach and Nowshera regions of Kashmir in 1917. In that year the Indian government asked the industrial chemist, William Raitt, to visit Kashmir and to report on the possibility of updating the craft-based paper industry in that region. Raitt was familiar with modern papermaking methods, which had already been introduced from Europe and were in operation in other parts of the sub-continent, and he had pioneered the use of bamboo as a source of papermaking fibre. By 1917 machine-made papers were already becoming available in Kashmir. Raitt's subsequent report concluded that there were no improvements to be made and that the ancient methods still in use would soon pass into history.

The final session of the conference began with Alan Crocker's study of The Ball family. Crocker had previously given a version of this paper at the IPH Congress in Vienna in 1992. Between 1763 and 1869 five generations of this family were papermakers at Rush Mill in Northampton, at Stoke, Chilworth, Albury Park and Postford Mills near Guildford, Surrey, at Gueres and Val Vernier Mills near Dieppe in Normandy, at St Sulpice Mill, near Doullens in Picardy and at Pont Audemer, also in Normandy. Little is known about Charles Ball I and his activities at Northampton, but Charles Ball II, of Guildford, was eminent as an inventor and as a manufacturer of superior banknote paper. In 1793-4 he made paper for the Comte d'Artois to produce forged assignats. His sons, Charles Ball III and Edmund Richard Ball, became papermakers at Postford Lower and Postford Upper Mills, but during the depression following the end of the Napoleonic wars, both became bankrupt. Subsequently Charles and his son Charles Ashby emigrated to France and, under the patronage of the Comte de Tocqueville, became great innovators. In particular, they were associated with the introduction of the papermaking machine into Normandy and with the development of new raw materials including straw. Finally Charles Ashby Ball took his son Allen Charles into partnership at Pont Audemer.

This was followed by Peter Bower's introduction to the complex subject of the Evolution and Development of 'Drawing' Papers and the effect of this development on Watercolour Artists between the years 1750-1850. For most of papermaking history papermakers had produced three basic types of paper: writings, printings and wrappings. 'Drawing' papers, which included papers for watercolour as well as those for pencil, chalk and ink, were quite simply those papers which artists found they could work on, regardless of the use for which papermills had actually made them. From the 1780s onwards, a complex web of technical, cultural and economic influences operated on papermakers, paper merchants, artists' colourmen, and artists themselves, which all led to the evolution of papers designed and made specifically for watercolour. The effects of the introduction of such papers coupled with new sensibilities and aspirations in the artists themselves had a dramatic effect on the working practices of individual artists and their

expression of their vision.

Watercolour, in particular, depends for its success on a deep and vivid understanding of the nature of particular surfaces and the ways that those surfaces can be worked. The ability of particular papers to satisfy the changing demands that artists now began to make on them was much influenced by the increasingly sophisticated technology of papermaking. The rapid introduction and exploitation of the paper machine by an increasing number of papermakers had made those who continued to make by hand change both their working practices and their concepts of papermaking itself, forcing them to produce papers of increased quality and range. They explored new raw materials and production methods, developed new qualities and characteristics; produced new surfaces, sizings, strengths, tones and colours. Watercolour artists seized on these new developments, not always using the new papers in the way the makers expected, but grateful for the opportunities they provided.

Much of this increasing specialisation was due to the activities of merchants and colourmen who were differentiating more and more subtle developments in paper qualities and characteristics as selling points when marketing individual papers, both to amateur and professional customers. Artists themselves contributed to this process, telling the colourmen what they wanted. The evolution of medium-specific papers depended on a rapidly growing understanding amongst makers of the possibilities inherent in their process and an equal growth of understanding amongst users of their own needs and requirements.

Many of these 'drawing' papers were primarily designed for technical drawing, rather than for the fine artist. The market for fine artists papers has always been relatively small, but by the early years of the nineteenth century the market for engineering, architectural, surveying, and map making was enormous. It was this market that made the development of such papers economically viable for the mills.

Heim Kropholler's paper, which followed, considered the link between paper type, or grade, and fibre characteristics. He showed that in the search for new fibre sources only small quantities of pulp are needed to characterise and estimate the papermaking potential of any particular fibre. Kropholler then considered the links between sheet properties, such as strength, air permeability and optical characteristics, and fibre traits, such as length, width, thickness, coarseness, shape and bonding potential, making particular reference to the potential role of straw. He outlined the three basic stages to be considered in such analysis: firstly, the basic fibre properties; secondly, the standard procedures that can be used to evaluate both handsheet properties and the effects of beating and refining on the fibres; and finally the manufacturing of a commercial grade of paper where considerations include other factors, such as fibre blends and the actual chemicals and types of equipment used.

The final paper of the conference came from Richard Hills, talking on James Watt and his Copying machine. The increasing correspondence involved in James Watt's expanding nation-wide business, supplying steam engines, led Watt to develop a method of copying all his letters onto a specially developed thin copy paper. Watt sent Matthew Boulton examples of his copying process in June 1779 and continued with his experiments until at least the November of the following year. In May 1780 he was granted a Patent and started to manufacture his copying machine. A batch of test samples are preserved in Watt's Garret Workshop in the Science Museum, London, and show that Watt had considerable problems with the paper,

trying many types from different makers. He wanted a thin, even, unsized paper which would retain sufficient strength when being handled wet during use. Watt very quickly turned to wove paper, one of the first commercial uses of such paper. While initial sales of the Copying Machine were disappointing, it was later to become an indispensable part of the Victorian Office. Recently more documents have been found in the James Watt Papers at Birmingham which give more information about the problems Watt had in procuring a suitable paper for use in his copying machine. These give details of different papermakers and show that several makers, previously not thought to be producing wove papers at such an early date, were producing them as early as 1780. These documents will form the basis of a further paper from Richard Hills.

This wide ranging programme, based as it was on the two loose themes of Straw and Oxford Papermaking History, produced much new historical evidence and some important insights into various developments in specific areas of this fascinating and complex industry. But equally, or perhaps more importantly, the whole conference was a very friendly and absorbing occasion.
Peter Bower

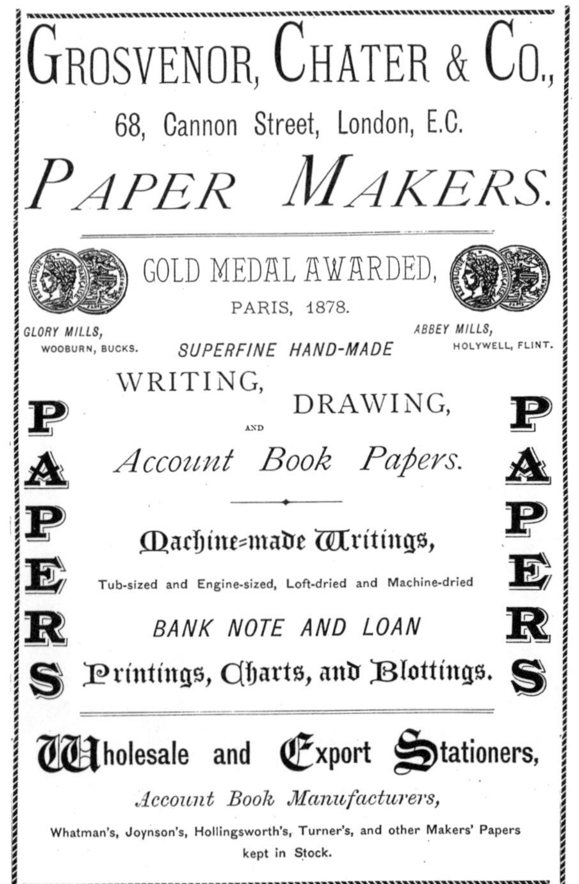

THE USE OF STRAW IN PAPERMAKING

Richard L. Hills

The earliest reference which I have discovered so far to the use of straw in papermaking has been in a French book about the Laroche family of papermakers who had mills at the town of La Couronne not far from Angoulême. Straw was tried around 1766 in the Limousin region because there was straw in abundance but a shortage of rags. However, because the resulting pulp was a brown colour, it was suitable only for making board or wrapping paper.[1] Presumably the straw was prepared in either stampers or the hollander beater without any chemical treatment. Already we see here some of the advantages and disadvantages of straw. Its advantage was its general availability in Western Europe although it was costly to transport through its bulky nature. Its colour has always remained a problem because bleaching it to make white paper has always been expensive. Because straw yields amongst the shortest cellulose fibres, the quality of paper made from it is generally poor and usually it has been mixed with stronger fibres. To what extent it was used, either on the Continent or even in Britain during the eighteenth century, I have no idea but I refer you to the work of Peter Bower on *Turner's Papers*.[2]

The more famous attempt to use straw in good quality paper was that by Matthias Koops when he set up a mill in London soon after 1800. The first edition of his book, *Historical Account of Paper*,[3] was printed in 1800 partly on paper made from straw. How much of the paper was actually made from straw and how much was actually made in England must be questioned but the frontispiece is printed on a yellow paper, presumably straw, which has survived in remarkably good condition although it seems the poorest of the three types of paper, straw, woodpulp or waste.

In August 1800, Koops applied for a patent for manufacturing paper from straw and other substances for which no specification was enrolled.[4] In the February of the following year, he applied for a further patent in which his processes were described.[5] He proposed to "manufacture paper from straw, hay, thistles, waste and refuse of hemp and flax, and different kinds of wood and bark, fit for printing and other useful purposes". The straw, hay or thistles were cut on a chaff cutting machine into two inch lengths, while the wood was reduced to shavings, again in two inch lengths. Sometimes these materials might be boiled for about three quarters of an hour or fermented for a few days to give them a preliminary softening. The basic process was to soak them in limewater for six or seven days, turning them so the lime was applied evenly. Then followed washing and boiling for a couple of hours in clean river water. Sometimes crystal of soda or potash might be added to improve the colour and texture. After this would be a further washing and boiling and then pressing to remove the water. Sometimes this pressed material might be fermented for several days or it could be made into paper in the usual way immediately, depending upon the type of paper needed.[6] Koops tried to establish the largest papermill in Britain at Millbank in London, but his creditors from an earlier bankruptcy tried to seize his assets in October 1802 among which was this mill. The mill itself was found to be heavily in debt and papermaking ceased there before the start of New Year, 1803.[7]

The reason for Koops's interest in making paper from these materials was the high price of the traditional linen rags during the Napoleonic Wars. Once trading conditions had returned to normal, interest in these alternative substances seems to have died down. This may have been through the greater use of chlorine bleach which became available in the 1790s and which presumably enabled the papermakers to make white paper from discoloured rags. It may also have been through the use of cotton as a furnish through the greater availability of cotton fibres as a result of the enormous expansion of that textile industry throughout this period. The early use of cotton is another area of papermaking which needs more investigation.

The increasing demand for paper throughout the nineteenth century put further pressure on the supply of raw materials. One source of supply, sails from ships, may have become scarcer as the century progressed due to the increasing number of steamships. It is also said that the change from rope hawsers to anchor chains for ships around the middle of the century caused a further diminution in the supply of suitable papermaking materials. Then the removal of the "Tax on Education", when William Gladstone abolished the final remaining duties on paper in 1861, is said to have stimulated a great increase in the demand for paper. While this may have been true, the pressure on supplies of raw materials for papermaking seems to have started both in England and on the Continent before this. Board and wrapping paper made from straw were being made in France from 1840 and from 1850 straw pulp could be bleached and used for newsprint.[8]

The British patents help us to see when inventors were interested in finding alternative sources of papermaking fibres. I looked through the *Patent Abridgements* relating to the manufacture of paper, pasteboard and papier mâché from 1665 to 1857 to find those concerned with pulping materials other than what I considered to be the traditional ones of flax, hemp and cotton. Some may have been missed because names of fibres were not included. The book finishes in 1857 so the figures are incomplete for the final decade. (See Table 1)

The increase in the number of patents in the decades 1831-40 and 1851-57 must be noted. The first may be linked to the lowering of the Excise duty in 1837 but the second certainly precedes the complete abolition of 1861.

The patents around 1800 are the ones taken out by Koops. The next patent concerned with straw was that of Louis Lambert in 1824 (does the name show Continental connections?). He patented a process similar to that of Koops and proposed to treat the straw:

> with quick lime or caustic potash, soda or ammonia... then use quick lime and sulphur to free it from the mucilaginous and silicious matter so prejudicial in papermaking... then wash it... bleach it with chlorine or exposure to the open air.[9]

There does not seem to have been any commercial exploitation of this patent and it is difficult to tell which patents were really significant. As well as patenting inventions, there was the lunatic fringe and it seems as if anything which might contain even a hint of cellulose fibres was at some time or another in the next few years patented as a source of papermaking fibres.

Looking at what was patented, we find patents for using plants which we might expect, such as "sugar cane after the sugar has been removed" in 1838,[10] or in 1854 Brazilian grass and Indian grass.[11] Again in 1838 there were suggestions for bark of trees, hop vine, stalks of potatoes and dried leaves,[12] or perhaps less pleasantly in the following year "horses' dung, and mixed with straw of any kind, in such a state as the straw is taken from the stables".[13] William Johnson produced a long list of plants all called by their Latin names in 1855,[14] while for paper that might have a more culinary flavour, James Acland suggested in 1854 "separating the fibrous portion of the roots of potato, parsnip, turnip, the roots, stems, & stalks of mangold wurzle, chicory, & rhubarb".[15] Horse radish and mustard were amongst other suggestions but the one I really liked was that of Josiah and Alfred Wright and Francis Roberts. They were to take the rhubarb stalk and use it not only to make paper but also wine and spirits. They stated that the stalk was to be taken

> at any age, passed between rollers to express the juice, the fibres are washed, etc., etc., and alone or otherwise made into paper. In making wine the juice is boiled, poured upon a certain quantity of lump sugar, fermented with yeast, strained, placed in a cask, etc. In making spirits the juice should be treated as in the first stage of making wine, and after fermentation distilled.[16]

It is a pity that they did not give more detailed recipes.

Alexander Watt, in his *Art of Paper-making*, claimed that the patent of J.T. Coupier and M.A.C. Mellier in 1852, was one of the more successful for using straw.[17] They reduced "vegetable matters into pulp by means of a solution of hydrate of soda or potash" and afterwards bleached it "by hypochlorites of alumina". Soda was the chemical most commonly employed for straw. To use straw, one of the problems was how to treat the whole of

<table>
<tr><td colspan="4">TABLE 1 — PATENTS CONCERNING FIBRES OTHER THAN LINEN, HEMP AND COTTON.</td></tr>
<tr><td>Years</td><td>Straw</td><td>Other fibres and straw</td><td>Total No. of papermaking patents</td></tr>
<tr><td>1665-1700</td><td>-</td><td>1</td><td>11</td></tr>
<tr><td>1700-1800</td><td>1</td><td>1</td><td>17</td></tr>
<tr><td>1801-1810</td><td>2</td><td>4</td><td>11</td></tr>
<tr><td>1811-1820</td><td>-</td><td>2</td><td>13</td></tr>
<tr><td>1821-1830</td><td>1</td><td>3</td><td>19</td></tr>
<tr><td>1831-1840</td><td>2</td><td>15</td><td>43</td></tr>
<tr><td>1841-1850</td><td>2</td><td>5</td><td>50</td></tr>
<tr><td>1851-1857</td><td><u>23</u></td><td><u>118</u></td><td><u>248</u></td></tr>
<tr><td></td><td>31</td><td>149</td><td>412</td></tr>
</table>

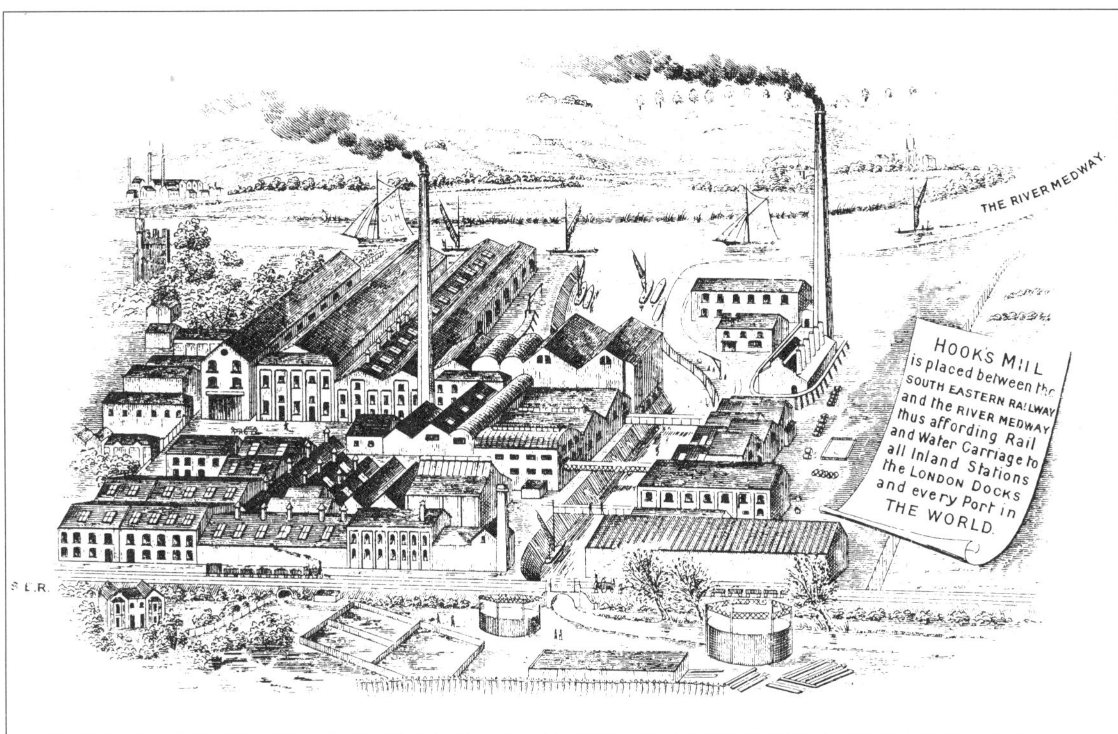

Figure 1: *Townsend Hook's Snodland Mill, circa 1886, from a drawing used on the verso of the company's business cards. The esparto and straw boiling plant is situated to the right of the creek.*

the plant equally because the nodes where the leaves branched off were denser than the rest of the stalk. Koops had cut his straw into two inch lengths with a chaff cutting machine while Lambert cut out the knots. George Stiff cut the straw up even shorter into half inch lengths and winnowed it to separate the knots,[18] while sometimes the straw might be crushed by passing through a mill or some other method of flattening before the chemical treatment.[19]

One of the next important developments was circulating the chemical liquor through the straw and not merely leaving the stems or stalks to soak in it. Presumably this not only ensured that the whole of the pulp was treated more equally but probably speeded up the process especially when the liquor was heated. We find these ideas in Charles Townsend Hook's patent in 1853.[20] After the straw had been cut up into lengths between one quarter and three quarters of an inch and the dust and knots removed, it was placed in an iron container 12 ft. diam. and 5 ft. deep which had a perforated false bottom through which the liquor could drain, leaving the straw behind. In the middle of the container was a standard which supported an arm which revolved and sprayed the liquor over the straw underneath it. The liquor was recirculated from a tank which had a steam heating coil to keep the temperature at boiling point.

Once again caustic soda or potash was the chemical used, sometimes followed by sulphuric acid in cold water which improved the colour. Townsend Hook found that straw took longer to bleach than rags.

This is an important patent because Townsend Hook continued to use straw for many years. Kelly's *Directory* of 1859 said of Hook's Snodland Mill, "Here is a manufactory of writing paper from straw which employes about 70 persons".[21] Eight years later, the *Directory* gave the work force as "about 120 persons". The production figures in Table 2 show the use and decline of straw there. In 1884, the fibre furnish for news print was 62% straw, 17% esparto, 9% woodpulp and 12% waste paper to which 16% of china clay was added.

The 1850s were a period when straw was thought to be an important source of papermaking fibre. On 7 July 1852, John Dickinson wrote to Charles Longman,

> Joynson has been making some paper with a large mixture of Straw and says it will answer well. It appears that it takes the size well, and handles extremely well, and thick.
> When it was tried at Bermondsey the knots were all cut out by hand. Might it be that these people cut it into Chaff, then make a

TABLE 2 — THE USE OF STRAW AT TOWNSEND HOOK'S SNODLAND MILL, IN COMPARISON WITH OTHER FIBRES [22]				
	1876	1884	1886	1894
Paper made, tons	2,252	3,762	4,167	4,840
Straw in £	10,327	17,248	9,313	17
	2,735 tons		3,866 tons	
Rags, esparto, etc.	10,797	17,622	24,936	589
Waste paper		196		1,508
Wood pulp				37,507

sort of half-stuff or long pulp and sift out the knots?

I have not been able to get a sight of Joynson's paper but from the description it is quite clear that it is not ill adapted for insides of all kinds of boards from Card Boards downwards... Rothney... says it was remarkably full in the hand and rattled uncommonly.[23]

John Dickinson does not appear to have used straw but his partner, John Evans, took out a patent for treating Brazilian grass in 1854.

In the third edition of his book published in 1863, Herring also saw a great future for straw and said that

> There is certainly no fibre to compete with it at present as an auxiliary to that of rags. A thick brown paper, of tolerable strength, may be made from it cheaply, but for printing or writing purposes only an inferior description can be produced, and of little comparative strength to that of rag paper. Its chief and best use is that of imparting stiffness to common newspaper.[24]

Cheapness of the raw material was one great advantage of straw. While Herring said that rags would cost at least £17 per ton, straw was only £2. However there would be much more waste with straw, that with rags being about one-third but with straw fully one-half. Then the cost of processing straw was much greater, two to three times that of rags, so that, in the end, "a similarity of value is thus ultimately attained.[25]

By this time, the straw was being boiled under pressure, which again would have reduced the time needed. Herring states that steam at pressures of 120 to 150 p.s.i. was being used. This may have originated in the patent of John Cowley in 1856.[26] He boiled the straw under pressure at a temperature of 250°F inside an iron vessel and continually circulated the liquor through it. In this way he could reduce the boiling time to between 8 to 10 hours. This is the first time these features appear in a patent and have remained in use ever since.

Another problem with straw as Herring pointed out was the smaller yield. John Balston gave a comparison of the cellulose yield in the dry state of various fibres, with cotton giving 91%, flax 82%, wood 60% and straw 40-50%.[27] These figures may have been derived from Cross & Bevan[28] where it can be seen that esparto gave a

Figure 2: *The Medal awarded to Townsend Hook at the International Exhibition of 1862. The* Illustrated London News *reported that "The webs of paper made from straw by Messrs. Townsend Hook & Co. are excellent, particularly so when made from that substance alone, unmixed with any other fibre — their printing and writing sorts being freer than usual from that brittle crackle so common to straw paper."*

roughly similar yield. However esparto was easier to treat because the nodes were absent, there was not so much loss and it did not need as much bleaching to give a good quality white paper. Straw was more highly lignified which was the cause of the difficulty in extracting the cellulose. It also gave a more brittle paper.

Watt in 1907 could say that "Very large quantities of this material [straw] are used in the manufacture of paper, but more especially for news-papers, the straw from wheat and oats being mostly employed.[29] Strawboard was also another important consumer of straw. Parkinson described *Straw Boards* as "a cheap material used by bookbinders as a substitute for millboards".[30] They were made from straw on a large scale "prepared from stable-yard matter". For straw boards, Hooper Davey said that the straw was packed between layers of lime and the whole treated by steam from between two to ten hours. Straw was seldom used by itself for making paper, partly through the shortness of its fibres, but was mixed with other pulps to give hardness. "For instance it will give a good "rattle" to a paper made with rag".[31]

During the First World War, many mills, especially those which could no longer receive imports of esparto, turned to straw as a raw material. One of them was St. Cuthberts at Wookey. Earlier experience with straw must have been long forgotten because when Edward Burgess took over the mill in 1856, he instituted the use of this material. John Sullivan, who managed the mill, said that his brother had seen samples of straw paper from France at the Great Exhibition of 1851. This mill was said to be the second in the United Kingdom where it was tried and their paper won honourable mention in class 28 for oat straw paper at the International Exhibition of 1862.[32] However this mill soon turned to esparto and continued to use this until a change to straw was necessitated around 1916.[33]

From February 1916, the import of papermakers' materials was more and more restricted and Bowaters at their Northfleet mill started to use straw.[34] Another mill was Dalmore on the Esk to the south of Edinburgh which used straw as a substitute for esparto between December 1917 and March 1919.[35] Guard Bridge[36] and Culter Paper Mills were others which had to make this change too. In their history, it is stated, "Culter and other mills of like equipment kept the wheels laboriously turning with Straw, but the result was a travesty of the normal product".[37] When peace was restored, straw was quickly abandoned as a material for good quality paper.

Perhaps that is the most appropriate note on which to end for, while many people have looked at straw as a pulp for paper, its disadvantages have always outweighed its advantages. I have not attempted a thorough survey of papermaking literature and I have been unable to find any figures to show the consumption of straw over the years, but I hope I have said enough to stimulate others to continue to investigate the history of this fibre.

References

1 Laroche, R. *Les Laroche papetiers Charentais*, Fumées du Nil, No. 23, Atelier Musée du Papier, Angoulême, 1992, p.54.
2 Bower, P. *Turner's Papers, A Study of the Manufacture, Selection and Use of his Drawing Papers 1787-1820*, Tate Gallery, London, 1990, pp.58 & 88.
3 Koops, M. *Historical Account of the Substances which have been used to Describe Events and Convey Ideas from the Earliest Date to the Introduction of Paper*, London, 1800.
4 Patent 2433, 1800.
5 Patent 2481, 1801.
6 Hills, R.L. *Papermaking in Britain, 1488-1988, A Short History*, Athlone Press, London, 1988, pp.133-4.
7 Hills, R.L. "Some Notes on the Matthias Koops Papers", *The Quarterly*, No.2, August 1990, pp.3-5.
8 Laroche, *op. cit.*, p.100.
9 Patent 5041, 1824.
10 Patent 7850, 1838.
11 Patents 964 and 979, 1854.
12 Patent 7794, 1838.
13 Patent 8138, 1839.
14 Patent 135, 1855.
15 Patent 1341, 1854.
16 Patent 1781, 1857.
17 Watt, A. *The Art of Papermaking*, Crosby Lockwood, London, 1907, p.16 and Patent 13,979, 1852.
18 Patent 393, 1853.
19 Patent 2319, 1853 and Patent 2278, 1854.
20 Patent 2133, 1853.
21 Funnell, W.J. *Snodland Paper Mill, C. Townsend Hook and Company from 1854*, C. Townsend Hook & Co., Snodland, c. 1979, p.26.
22 *ibid*, pp.42 & 43.
23 Evans, J. *The Endless Web, John Dickinson & Co. Ltd., 1804-1954*, J. Cape, London, 1955, p.110.
24 Herring, R. *Paper and Paper Making, Ancient and Modern*, Longman, Green, London, 3 edn. 1863, p.67.
25 *ibid*, p.68.
26 Patent 1039, 1856.
27 Balston, J.N. *The Elder James Whatman, England's Greatest Paper Maker, 1702-1759*, J.N. Balston, West Farleigh, 1992, Vol II, p.188. note.
28 Cross, C.F. & Bevan, E.J., *A Text-book of Paper-making*, E. & F.N. Spon, London, 4 edn. 1916, pp.100 f.
29 Watt, *op. cit.*, p.16.
30 Parkinson, R. *A Treatise on Paper*, R. Parkinson, Clitheroe, 3 edn. 1894, p.144.
31 Hooper Davey W., *The Romance of Paper*, Birmingham, 1912, p.27.
32 Luker, B.G. *Mill 364, Paper Making at St. Cuthberts*, St. Cuthberts Paper Mill, Wookey, 1991, p.13.
33 *ibid*, p.35.
34 Reader, W.J. *Bowater, A History*, Cambridge University Press, 1918, p.20.
35 Watson, N. *The Last Mill on the Esk, 150 Years of Progress*, Scottish Academic Press, Edinburgh, 1987, p.85.
36 Weatherill, L. *One Hundred Years of Papermaking, An Illustrated History of the Guard Bridge Paper Company Ltd., 1875-1975*, Guard Bridge Paper Co., Guardbridge, 1975, pp.20 & 79.
37 Anon, *History of Culter Paper Mills, Two Hundred Years of Progress*, Aberdeen, c. 1951, p.32.

BY ROYAL LETTERS PATENT
Queenhithe HARRY A. BARTON Upper Thames St
LONDON
Patentee & Sole Importer
PATENT SOFTENED
VEGETABLE PARCHMENT
In Rolls,
Sheets, Strips,
or Squares for covering
Jars and Packing purposes
Being
Non-porous, it is
specially adapted for
protecting articles from the atmosphere
ORDINARY VEGETABLE PARCHMENT
STEARINE, PAPER AND TINFOIL.
SAMPLES POST FREE

STRAW IN 19th CENTURY PAPERMAKING

Peter Bower

For most of the past century or more of paper historical study, paper historians have had little interest in the actual physical make-up of individual sheets of paper, preferring to explore the world of watermarks found in white papers and to document the histories of individual mills. More recently the industrial, technological, economic and social histories of the industry have received increasing attention. But one aspect of paper history that, except in very general terms, has had little attention is the nature and usage of specific fibres and in particular how individual fibres have been variously developed to achieve very specific results.

This paper does not pretend to be comprehensive, being more an introduction to some of the history of one type of fibre: cereal straws. Richard Hills, in his paper, has covered some of the technical developments as found in patent information. Together these two papers will hopefully suggest further areas of research into the use of straw, and other alternatives to rag, in western paper-making that might profitably be explored.

Straw has played a part in European papermaking since its very early years, but it has never been a major part. Microscopic examination of papers from before Matthias Koops' well documented manufacture of pure straw papers in 1800[1] has shown that its primary function with one or two exceptions was, prior to that date, as a bulking fibre rather than a primary raw material. It was cheaper than linen or hemp and at some times of the year, readily available locally to most mills.

The written evidence for the use of straw in papermaking before 1800 is almost non-existent, but evidence as to its use does exist in the papers themselves. However compiling this evidence is a very time-consuming process as it involves the microscopic examination of individual sheets of paper where the use of straw is suspected and much more work remains to be done on this.

From such examinations it appears that the earliest usage involves the use of various straws as fillers in combination with either low grade linen rag or low grade linen and old hemp rope mixtures in the production of brown, whited-brown and some blue wrappings. Quite simply a load of roughly chopped straw was added to the rags in the stamper or beater fairly late on in the beat. The straw was being used unbleached and basically just broken up rather than actually beaten. It is often easily visible as larger particles, visually similar to the specks and shives often found in white papers. This is very different to the

way straw was to be used later in the nineteenth and twentieth centuries, where the natural 'wetness' of straw fibre in beating was exploited to produce low bulk and a certain harshness or crispness in the sheet, character-istics that were particularly useful in *banks, loans* and *bonds* where a crisp handle and firmness in the sheet were considered of great importance.

Although straw was being used as one constituent of some papers made from more than one fibre in western Europe from the 16th century onwards, there is no evidence of it being used as the main or sole ingredient of any paper, other than in some pasteboards from the 16th century or in Schäffer's experiments[2], until Koops' productions of 1800.

I have so far found no evidence of it being cooked or deliberately bleached, or of a sufficient beating that achieved any real fibrillation before the beginning of the 19th century. What one sees under the microscope is something used to bulk out other more expensive materials. Straw only remained a cheap filler as long as there was no cooking, chemical treatment, washing or true beating involved. As soon as straw began to be seriously considered as the sole or principal raw material it proved to be a difficult fibre to work. Its short fibre length, its colour and the presence of silica and lignin, in higher quantities than are found in many other plants, have continuously led to production problems.

Distinctions must be made between the various types of straws, e.g. wheat, oat, corn. Microscopically the most striking features of the cereal straws are the very large, thin walled, pith cells and the serrated cells from the outer wall of the stem. The fibres are from 0.5mm to 2.5mm in length and 8μ to 20μ wide. Cereal straws have a dry-basis content of 31-40% cellulose, 35-48% hemicel-lulose and a 15-25% lignin content. Straw has a higher mineral content (4-8%) than other fibres. The chemical combination of one variety of straw from the same field varies from year to year, dependent on the weather and growing conditions.

During the nineteenth century many attempts, some very successful, were made to find economic alternative fibres that the new technologies could develop into papers of very fine qualities. The two main rivals besides wood as an alternative fibre for paper appear, at least in Britain, to have been cereal straws and esparto grass, including related fibres such as alfa and diss. Esparto won out because of the relative ease of its processing in

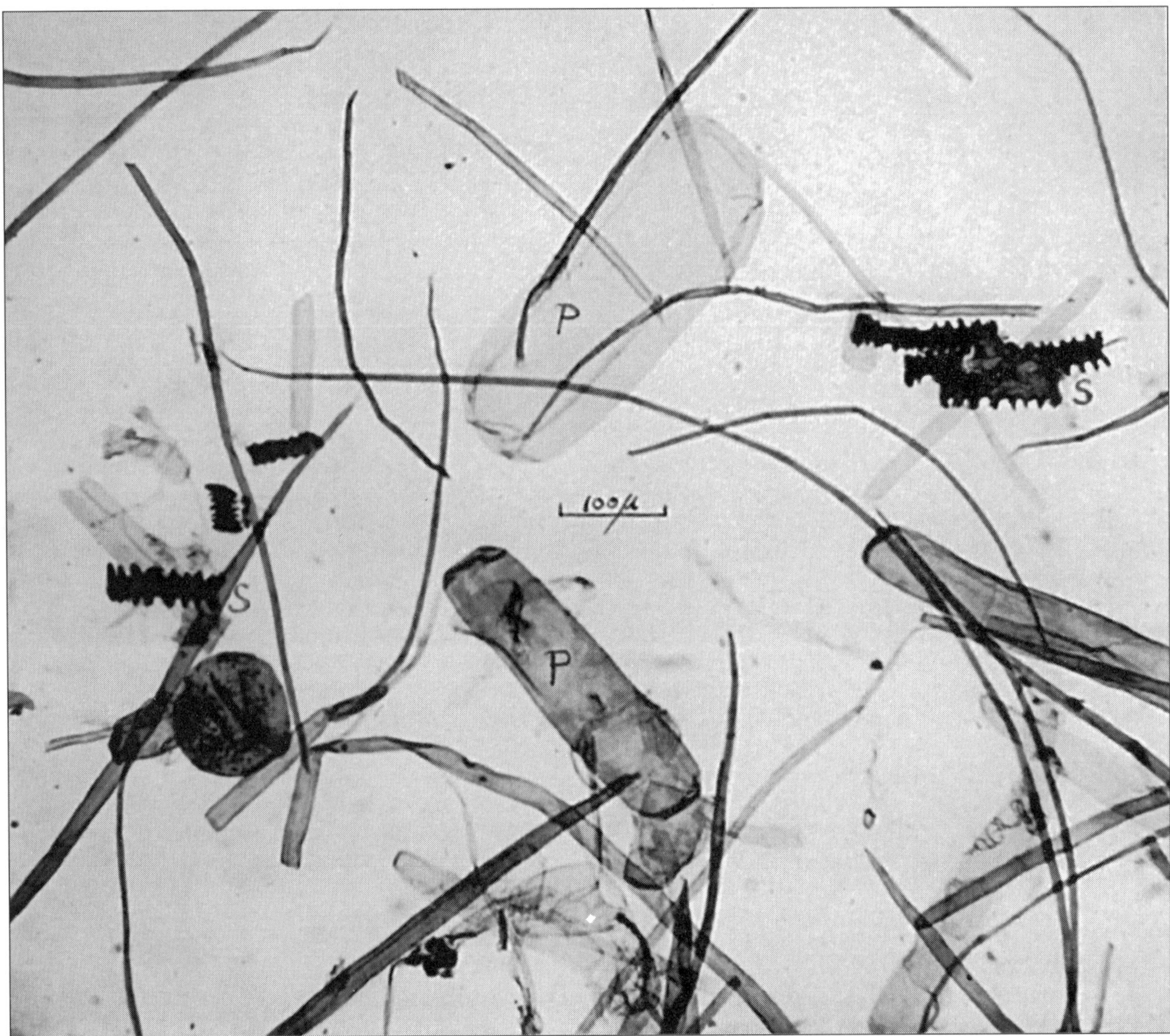

Figure 1: *Photomicrograph of a cereal straw, in this case wheat, showing the characteristic large, thin-walled, pith cells (marked P) and the serrated epidermal (outer wall of the stem) cells, marked S. Contrast the cell forms and sizes with those seen in Flax (Figure 2). (Armstrong)*

comparison with straw, and the ease of its availability, despite it having to be imported from Spain and North Africa.

As Richard Hills has pointed out[3], straw has always been considered as very widely available, indeed some have said that it is perhaps the most widely available waste product that has a use in papermaking. But one of the problems with straw is that despite its apparent abundance, not that much is often readily available to the papermaker. The waste that farmers don't keep for their own use is often too expensive to collect and process. Despite the massive increase in understanding of the science of papermaking and in technological innovation, particularly in pulp preparation, over the years, straw is still not an easy fibre to work.

Esparto won out over straw, at least in nineteenth century Britain. Some of esparto's advantages over straw were as follows:

1. Its small thickness offered little resistance to liquid penetration.
2. Its structure, without knots or seed hairs, gave a more uniform pulp.
3. Its chemical composition as a pecto-cellulose made resolution possible at low steam pressure or even at atmospheric pressure.
4. The yield of fibrous material was higher.[4]

Figure 2: *Photomicrograph of Linen, Linum usitatissimum (Flax), fibres after 6 hours in the beater, showing very characteristic fibrillation and the much greater fibre length than is seen in straw.* (Armstrong)

Routledge considered that as regards esparto:

> There was nothing in the range of fibres that could touch it, even at the present price, for the purpose of papermaking, considering either quality or value. He had seen most of the fibres that came into England, and none could compete with it practically.[5]

The nineteenth century was to see much discussion of a wide variety of alternative fibres. This paper will concentrate on people's own words to describe how they saw the possibilities and pitfalls of straw as a papermaking raw material in the nineteenth century. Rees in 1819, presumably thinking of Matthias Koops' experimental production, says that

> Trials of straw have been made in England as far back as the year 1799; but much remains to be done; and as the occasion is important, especially as it relates to commerce ... we should be happy to find, that some public spirited individual, of enterprise and talent, would turn his attention to an extensive series of experiments; for the purpose of ascertaining the merits of these or other substitutes for linen rags, in the fabrication of paper for writing and printing.[6]

There were in fact many such "public spirited individuals of enterprise and talent" who spent long hours on both trials of alternative fibres and on disseminating their ideas and processes. The publisher John Murray had some paper made from straw but in 1824 was personally more convinced of the merits of nettles, which

> would be an excellent substitute for linen rags, if linen cannot be obtained in sufficient quantity. At Nisa, in the North of Italy, they manufacture a beautiful cloth from the parychymatous fibre of the nettle. ... I anticipate having soon an opportunity of putting the suggestion of the nettle to the test of experiment through the medium of a paper manufacturer.[7]

Unfortunately it is not known who Murray approached for this trial or what their results were, or even if the projected trials ever took place. Some of the experiments being conducted into straw, and other fibres, did however find their way into print: Mr. Lambert's "Process for making paper of Straw"[8] is described in great detail in *The Engineers and Mechanics Encyclopaedia*[9] only a few years after another encyclopaedia had written

> Paper has been occasionally made of straw, and other materials not commonly in use, and a Mr. Koops, in 1802, obtained a patent for making straw-paper, but we have not heard that the use of this article is become common.[10]

The Society of Arts proved to be a very important forum for such discussions, holding many meetings on the subject of papermaking fibres, attended by papermakers, scientists and interested amateurs. Robert Collyer was moved to declare that

> No one with whom I have discussed the question denies the proposition that the moment the silica is removed from straw they will produce equally as good and no doubt a much stronger paper than rags.[11]

The discussion was not just confined to Britain: Justus v Leibig had written to Collyer, from Munich, in 1859 saying

> The communication regarding your important discovery of converting the fibre of straw into a pulp, capable of being made into all kinds of paper, has received my attention. It gave me true pleasure, because the substitution of a suitable material for rags in the manufacture of paper, cannot be regarded as otherwise than highly desirable.
>
> I am moreover of the opinion that the simplicity of your process, as well as its cheapness of application, renders your beautiful discovery highly valuable, and I can only wish that it may be duly appreciated as it deserves to be, and that you may recoup rich fruits from your labours and experiments.[12]

Not everyone was so optimistic about the actual realities of applying patents and processes such as Collyer's to practical paper making. Mr. Norris said that

> He knew something of agricultural pursuits, and he could say that in all the Midland Counties, and in most of the corn growing districts, straw not only did not abound, but for papermaking purposes there was a strong prejudice on the part of the agriculturists against parting with their straw, and if they did part with it it was only done in an excited market. ... It was within his own knowledge that several mills in Gloucestershire and Oxfordshire, which were devoted to the manufacture of paper from straw had proved lamentable failures, owing to the inability to obtain the straw at prices which left a remuneration to the manufacturer. They were all aware that the spirit of enterprise amongst farmers of the present day was such that they did not depend upon their crops of corn for their profits so much as upon the fattening of cattle for the meat market. Hence the necessity of keeping the straw produce of their farms for their own purposes.[13]

Collyer finished the discussion by pointing out that

> 30,000,000 tons of straw were produced annually as the refuse of the necessaries of existence - wheat, barley and oats. There could be no difficulty if it were used, as it might be used, without the mixture of other substances and paper would be made at one-third of the price at which it was made from rags. This was no speculative assertion. It was being done in America, where Mr. Nixon, near Philadelphia, was making 40,000 tons of paper per annum from straw, and realising £10,000 a year profit.[14]

In another paper given to the Society of Arts, P.L. Simmonds discussed the possibilities of wood, esparto, alfa, plantain, yucca, bamboo, baobab, screw pine, dwarf palm, New Zealand flax and the fibrous husk of the betel nut.[15] He also quotes a meeting of the Scottish Papermakers' Association which had met to discuss a home-grown alternative fibre and which was also well aware of the actual problems of supply as opposed to the theoretical abundance of straw:

> The general opinion was that straw was the cheapest and best material that could be had, but great difficulty exists in obtaining large supplies, owing to the lease stipulations [on Scottish farms at least] requiring it to be consumed on the farm. The market price of straw here being also £6 per ton above what it is in Sweden and Belgium, it can be imported from those countries cheaper than it can be bought here.[16]

Despite all the problems, straw had become a common constituent of many 'rag' papers by 1860, although as Julius Grant describes it

> one might more correctly regard it as an adulterant of rag, because owing to the difficulty of processing straw in those days, the resulting pulp was dirty and had a poor colour. At this time less than 10% of the total output of paper and boards in Great Britain was made from Straw, and most of this was used for low-grade papers.[17]

Throughout the middle of the 19th century the *Journal of the Society of Arts* records a wide range of papers and correspondence discussing the use of straw and other plant alternatives to rag. Some of these are most illuminating, but must be treated with some caution. Many of the contributors, though learned and honourable men, were not papermakers and were relating their information at second hand. Some indeed seem to have approached the subject with more enthusiasm than sense. One particularly fascinating and unintentionally humorous contribution is that of George Lloyd regarding the use of Cow Dung, where the author describes the cow as

> a natural and most economical "rag-engine" for the separation and comminution of the fibre in the jaws and teeth of the ruminant machine - a series of macerating vessels in the stomachs and alimentary canal in which the soluble matters are detached and removed, not as waste, but destined not only to keep in repair the "machine" itself, but by increase of weight add most materially to its value.[18]

The arguments about the actual, as opposed to theoretical, availability of large amounts of straw for papermaking went on and on. An unidentified correspondent in the *Journal of the Society of Arts*, signing himself only "Waima", having discussed the price and suitability of various fibres for papermaking, has the following to say:

> I am of the opinion that an unlimited supply of a cheap and suitable material exists in our own country. I refer to straw. The sheet upon which I write is made entirely from straw, and leaves little to be desired for ordinary uses, and for many purposes it is preferable to paper made from rags. Moreover, less power is required to prepare the material, the process being more chemical than mechanical, an important matter when the high price of coals in some parts of the country is considered. Why, then, has this manufacture been comparatively neglected? Solely, I believe, from the circumstance that the large quantity of alkali required to prepare straw for pulp, by combining with its resinous and siliceous matters, causes that article, the alkali, to become a more important element of cost in the manufacture than the straw itself.[19]

The writer then goes on to describe various ways in which the cost of such processing might be reduced by the resale of the resulting waste for use "as the raw material of some other manufactures, such as soap making, or for common glass" and suggests that

> The proprietors of the following straw-paper mills, I believe all at present in existence, would, I have no doubt, supply some of their "black liquor" to any soapmaker, glass manufacturer or chemist who might be disposed to try experiments with it, viz:- Tovil Mills, Maidstone, Kent; Quenington Mills, Fairford, Gloucestershire; Burnside Mills, Kendal, Westmorland; Golden Bridge Mills near Dublin.[20]

The nineteenth century, perhaps more than any other previous period of papermaking history was affected by technological developments resulting from efforts to meet the ever-increasing demands for more and cheaper paper. One interesting survey of 19th century papers that provides us with some insight into how much and to what effect straw was used can be found in William Barrow's researches into the permanence and durability of book papers. His laboratory examined 500 19th century books, 50 from each decade, choosing

> in general little used non-fiction works showing no visible evidence of heavy use, abuse, mould or storage under unusual conditions.[21]

Barrow shows that in terms of the fibres used for book papers, at least in America, the nineteenth century can be divided into three distinct periods: 1800-49 when the papers were entirely rag based; 1850-69 when the fibre ranges from rag to chemical wood; and 1870-99 where rag, chemical wood and straw are all used. He found that "in the final decade of the century all rag disappears entirely from the sample"[22]: 80% are part rag and 20% contain no rag at all.

The books Barrow used were primarily American and thus do not reflect either British or European practice and usage as regards straw, nevertheless, as his is the only such survey I have come across, it is perhaps worth considering.[23] According to Hunter, paper using straw had been made by William Magaw, at Meadville, Pennsylvania and also at Chambersburg, Pennsylvania as early as 1828.[24]

Barrow found no papers containing straw amongst his 500 examples prior to 1853 when one book contained a paper with a 95% rag 5% straw furnish. In the following decade he found 3 examples; in the 1870s, 15; in the 1880s, 8; and in the 1890s a further 9. The table listed in the Appendix shows some vague patterns of 19th century usage particularly with reference to the blending of straw with other fibres. The highly specific nature of Barrow's researches, into the permanence and durability of book papers, determined to a great extent his choice of material to examine. It is highly likely that if works of fiction had been included then the data would have included some very different results with a much greater usage of straw in low grade printing papers.

Production problems plagued the straw papermakers. Over the latter part of the nineteenth century the convention grew up that straw was only used mixed with other fibres. By 1884 it was stated that

> A considerable quantity of straw is used both in Britain and America for papermaking. In general it is mixed either with rags or with esparto, being of too brittle a nature when bleached to make into paper alone. ... As at present treated, the yield averaging only 33-40%, straw will not come into general use, except in cases where the raw material can be bought on unusually advantageous terms. There is no doubt that in this case especially, a more rational method of extracting the cellulose, than by boiling under high pressure with a large amount of caustic soda is most desirable, for, many of the fibres of straw being extremely fine, these are to a considerable extent actually dissolved by the soda, and, whereas theoretically straw with 15% moisture ought to produce 45% cellulose, by the soda treatment not more than 33% are obtained, where a good white colour is needed.[25]

But as the century ended straw had become a regular part of various types of papermaking in Britain and many other countries:

> Straw was used a century ago for papermaking, but its extensive use is of comparatively recent date. For low papers it commands a market, but as a mixer it is inferior to Esparto. ... Over 50,000 tons of straw are imported from Germany, Holland, Belgium etc.[26]

As usage increased so did the specialist equipment: in 1876 James Bertram & Son were advertising machines for "Labousse's straw-stuff process for white papers."[27] Merchants such as J.A. Reid of Leadenhall Street, London were advertising "Bleached or unbleached Straw half-stuff".[28] It is a little difficult as yet to ascertain exactly how many British mills were producing and using straw pulp by the end of the century but some figures are available for both France and Germany. Bryan's *Directory* lists nine mills that produced straw pulp for sale to the French paper industry[29] and a staggering 52 German "straw pulp manufacturers".[30]

The twentieth century has seen many changes and fluctuations in the use of straw as a papermaking fibre:

> Before 1939, straw was used fairly widely outside this country for making paper, but it became the only effective and plentifully available substitute for esparto when the importation of esparto was cut off during the war period. Consequently much has been learned about the preparation of straw fibre, and it is still of considerable use,[31]

and anyone wishing to follow the changes in both processing and usage of the material is recommended to consult the various textbooks which discuss this fibre and its processing.[32] Although given little thought outside some parts of the paper industry, straw has been in almost continuous use as a production fibre in western European papermaking for nearly two hundred years, either as an unbleached pulp for strawboards and low grade wrappings, or as bleached cellulose for finer papers. For example, the wide ranging occurrence of straw in particular types of paper, in production in Britain in the mid 1950s, was noted by Armitage in 1957 as follows:

Antique wove and laid, Art: Body Papers, Azure Laid, Banks and Bonds, Offset Cartridge, Cheque papers, Chromo: base papers, Collotype, Cover papers, Cream wove or laid, Drawing paper, Duplicator paper, Imitation Art, Index cards, Ivory Boards, Litho and Music papers, Machine coated papers, Map papers, Middles and Tickets, Parchment Wove, Pasteboards, Pastings, Printings, Pulpboard, Strawboard, Straw paper, Ticket boards.[33]

Although the use of Straw in fine papers is a rather different process than making strawboard I will end with a description dating from 1919, by P.J. Machin, a leading strawboard manufacturer, of one form of Dutch straw board processing that for its simplicity cannot be beaten:

At one end of that machine there is the straw in a loft; the straw comes down a chute and it comes under a grinding stone and that crushes it: it comes through under knives, and they chop it up; it falls through into a trough, and there they swill it, and put some glue in it; it follows from there into a machine and goes through 20-25 rollers, and without a mortal man touching it, it comes out at the other end, lined strawboard, all ready dried for the market.[34]

Machin was giving evidence to the 1919 *Paper Industries Inquiry Committee*, under the chairmanship of H.A. Vernet, set up to "find out the causes which have led to the closing down of mills in England and to unemployment", but despite the closures and some of the difficulties of working with straw, he added later in his testimony

All I can tell you gentlemen is, that I do not know a strawboard proprietor who is poor; they all seem to make money out of it.[35]

References

1   Koops, Matthias, *Historical Account of the Substances which have been used to describe Events and Convey Ideas from the Earliest Date to the Introduction of Paper*, London, 1800.

2   Paper made from straw, but with a small admixture of linen rags can be found in volume II of Schäffer, Jacob Christian, *Versuche und Muster Ohne alle Lumpen oder doch nit einen geringen Zusatze derselben Papier zu machen*, 6 volumes, Regensberg, 1765-71.

3   Hills, Richard, *The Use of Straw in Papermaking*, in this volume, pages 9 – 13.

4   see Anon, *Esparto Paper*, London, 1956, part 2, p.3.

5   Quoted in the report of the Discussion that followed Robert Johnston's paper, at The Society of Arts, on *Esparto: A Series of Practical remarks on the Nature, Cultivation, Past History and Future Prospects of the Plant; including a Demonstration of the Importance to the Paper-Making Trade of Prompt and Vigorous Measures for its Preservation*, published in the *Journal of The Society of Arts*, December 22 1871.

6   Rees, Abraham, "Paper" in *The Cyclopedia or Universal Dictionary*, London 1819.

7   Murray, John, *Observations and Experiments on the bad composition of Modern Paper*, London, 1824.

8   Louis Lambert, Patent No. 5041, 23rd November 1824.

9   *Engineers and Mechanics Encyclopaedia*, London, 1836, p.235.

10   *The Book of English Trades and Library of The Useful Arts*, London, 1824, p.292.

11   Collyer, Robert, *Journal of the Society of Arts*, April 27th 1860, p.428.

12   Leibig, Justus v., Letter to Collyer in *Journal of the Society of Arts*, May 11th 1860, p.514.

13   Norris, J.T., MP, in the discussion following Robert Collyer's paper "On Paper Material", *Journal of the Society of Arts*, April 20th 1860, p.404.

14   Collyer, Robert, *ibid*, p.404.

15   Simmonds, P.L., in the *Journal of The Society of Arts*, January 27th 1871, pp.171 ff. Another interesting document on this subject is the researches of Dr. Mueller, of Melbourne, Australia, into "The Barks, Foliage, Grasses and Rushes of Australia suited for Paper making" in the *Journal of Applied Science*, August 1870.

16   Simmonds, *ibid*, p.175.

17   Grant, Julius,, *Wood Pulp*, Leiden, Holland, 1938, p.18.

18   Lloyd, George, "Manufacture of Paper from Cow-Dung", in the *Journal of the Society of Arts*, 1853, pp.368-9.

19   anon, "Materials for Papermaking", *Journal of the Society of Arts*, 1854, p.404.

20   *ibid*, p.404. The makers operating these named mills at this date are as follows: Tovil Mills, Kent - Allnutt and Son (see Bradshaw's *List of Paper Mills in England, Scotland and Ireland*, 1853), Quenington Mill, Gloucestershire - owned by J. Cowley and managed by C. Brewer, (see Harris, F.J.T., "Paper and Board Mills in or near Gloucestershire" *Transactions of the Bristol and Gloucestershire Archaeological Society*, 1976, Volume XCIV, p.135), Burnside Mill, Cumbria - James Cropper (see Bradshaw, *op. cit.*), Golden Bridge Mill, Dublin - D. & J. Sullivan (see Bradshaw, *op. cit.*).

21   Barrow, William J. (ed), *Strength and Other Characteristics of Book Papers 1800-1899*, Volume V of *Permanence/Durability of the Book*, Richmond, Virginia, 1967, p.9.

22   *ibid*, p.26.

23   The one English book that contained straw was W. Hook's *The Lives of The Archbishops*, published in London in 1879 (see Appendix: ref. No.510, below).

24   Hunter, Dard, *Papermaking*, New York, 1947, p.545.

25   "Paper", in *Encyclopaedia Britannica*, 9th Edition, 1884.

26   "Paper" in *Chambers Encyclopaedia*, London, 1901, p.739.

27   Craig's *Papermakers Directory Advertiser*, London, 1876, p.17.

28   *ibid*, p.19.

29   Clark W. Bryan & Co., *The Paper Mill Directory of the World*, Holyoke, Massachusetts, 1883, pp.105-6.

30   *ibid*, pp.130-132.

31   Norris, F.H., *Paper & Papermaking*, London, 1952, p.10.

32   e.g.: Bolam, F.M. (ed), *Stuff Preparation for Paper and Paperboard Making*, Oxford, 1965. Carpenter, Charles H. and Lawrence Leney, *91 Papermaking Fibers*, New York, 1952, Clapperton, R.H. & William Henderson, *Modern Papermaking*, Oxford, 1941, Emerton, H.W., *Fundamentals of the Beating Process*, Kenley, 1957, Hardman, H. & E.J. Cole, *Paper-making Practice*, Manchester, 1960, Tucker, E.C. et al, *The Manufacture of Pulp and Paper*, Volume IV, New York, 1924, and the three editions of the Technical Section of the British Paper and Board Makers's *Papermaking*, Kenley, 1949, London, 1965, London, 1978.

33   Armitage, F.D., *An Atlas of the Commoner Paper Making Fibres*, London, 1957, pp.147-152.

34   Captain W.E. Nuttall's MS copy of the records of the *Evidence given to the Paper Industries Inquiry Committee*, Wednesday 16th April 1919, p.120. Nuttall was a member of the Committee.

35   *ibid*, p.124.

APPENDIX: *19th Century Book Papers with a Straw Content*

The following information was abstracted from Barrow, William J. (ed), *Strength and Other Characteristics of Book Papers 1800-1899*, Volume V of *Permanence/Durability of the Book*, Richmond, Virginia, 1967. Appendix B, Table 1 and the complete list of 500 titles and their publication data. Anyone wishing to know which particular books were examined is referred to the publication itself. Papers with a more than 50% straw content are marked *.

Barrow's Ref. No.	Date of Publication	Where Published	Fibre content (in %):				
			Rag	Chem. Wood	Hard Wood	Soft Wood	Straw
834	1853	New York	95				5
813	1865	New York	60		20	10	10
587	1867	New York	90	5			5
588	1869	New York	90	5			5
644	1870	New York			20	60	20
508	1872	New York	90				10
646	1872	New York			40	40	20
702	1872	Philadelphia			65	10	25
645	1872	New York			30	50	20
594	1872	New York	30		20	20	30
647	1872	New York			30	50	20
708	1873	New York			10	80	10
798	1873	New York	50		5	5	40
790	1873	Boston			10	20	70*
641	1875	New York			20	60	20
709	1876	Philadelphia			10	70	20
706	1878	Cincinatti			60	30	10
707	1879	New York			30	60	10
510	1879	London	90				10
605	1880	New York			30	50	20
632	1881	New York			60	10	30
602	1882	New York			60	20	20
770	1884	Boston	10		50	20	20
637	1886	New York			60	30	10
694	1886	Philadelphia			70	20	10
606	1887	Boston			60	20	20
609	1887	New York			10	10	80*
758	1890	New York	10		50	30	10
616	1890	Philadelphia			45	30	25
617	1891	New York	10		25	20	45
759	1896	New York			20	40	40
620	1897	New York	10		50	30	10
742	1898	New York	10		50	30	10
760	1898	New York			10	60	30
757	1898	Boston	5		55	30	10
756	1899	New York	10		30	40	20

PAPERMAKING IN THE OXFORD AREA

Frances Wakeman

Papermaking began in the Oxford area late in the 17th century and at various times over twelve mills were dispersed throughout the County (pre-1947 boundaries). Papermaking started at South Hinksey in 1675, at Wolvercote by 1678, Hampton Gay in 1681 and at Eynsham about 1682. The makers of white paper in the early days had an advantage from the liberty of Oxford University to print and license printing in its precincts. The presses in Oxford used substantial quantities of paper from local mills and in 1855 the University bought the mill at Wolvercote. Today only the mill at Wolvercote survives.

The decline of the mills in this area was due mainly to its remoteness from the coalfields which increased their operating costs. Sandford mill was the only one with enough water power to drive a Fourdrinier machine. Another factor was the regulations imposed by the Thames Conservancy about the discharge of waste waters, far stricter than in other industrialised regions.[1]

In this short survey the mills of Hampton Gay, Eynsham and Sandford are included together with the mills in the north of the County.

Hampton Gay

Hampton Gay is a picturesque deserted village on the River Cherwell six miles north of Oxford. All that remains of the paper mill today are the waterways with some gear in the headwater and part of the metal breast shot wheel.

In 1681 the Manor was owned by the Barry family and Vincent Barry leased the grist mill to John Allen at a rent of £9 p.a. to use only as a paper mill.[2] There were some changes of owner but Thomas Miles occupied the mill in 1761 and later John Miles.

There were five two-vat mills in the County at this time[3] but there is no record as yet of the number of vats at Hampton or of the quality of its paper. From 1815-6 the Manor House, the farm and the mill were rented by Charles Venables, a member of the well known paper-making family at Cookham[4] A 48 inch Fourdrinier machine was installed in 1819 for £1,000 plus installation costs.[5] From the Baptisms recorded in the Parish Registers at least 8 families were listed with the father as papermaker, apart from the Venables. The 1841 Census records 17 houses with a total of 74 persons and 5 men listed as papermakers.

From the records at the O.U.P. Venables was supplying paper to the Bible Press from 1818 onwards and to the University Press between 1838-51 when he retired. The sums involved varied from £4,000-5,000 p.a. to the Bible Press and much lesser amounts between £500 and £1,000 to the University Press.

In 1849 the property was advertised for sale and the mill was described 'as a substantially built paper mill with various machinery and 12 acres. Two water wheels turned by a powerful stream, six iron engines, three rag boilers, three Screw presses ...A Machine House, Wheel House, Bleaching House, Steep Room, Rag Room and Drying Lofts.'[6]

Venables bought the property including the Manor House and farm, but it is debatable as to how active he was in the papermaking, as there were several changes of tenant. In 1863 the property was sold to Wadham College for £17,500 and the mill was leased to James Durham at an annual rent of £268 1s 6d.[7]

The mill was reconstructed between 1863-73 and a gas works, a steam engine, boiler and other machinery were installed but the new works were destroyed by fire in 1875.[8] R. Langton Pearson was the tenant at the time and an interesting side issue was a dreadful railway accident on the nearby Oxford - Birmingham line when the dead were taken into the mill.[9]

By 1880 the mill was operational again with the machinery in good working order and production was a ton of paper a day, mainly Browns and Groceries.

Despite all the improvements, there were several bankruptcies and the last tenants, J. & B. New, sold their stock in trade for the benefit of their creditors.[10] Hampton Gay suffered from its isolation as there was no metalled road, the only link was by canal to the coalfields of the Midlands and to Oxford. In the church there is a monument to two of the Venables children and Sarah, his wife.

Eynsham Mill

The remains of Eynsham mill are on the River Evenlode, near its confluence with the Thames, about five miles above Oxford. Today only the Mill House remains and the site is a fish farm.

A corn mill was recorded here in Domesday Book and in 1682 the mill was converted to white papermaking.

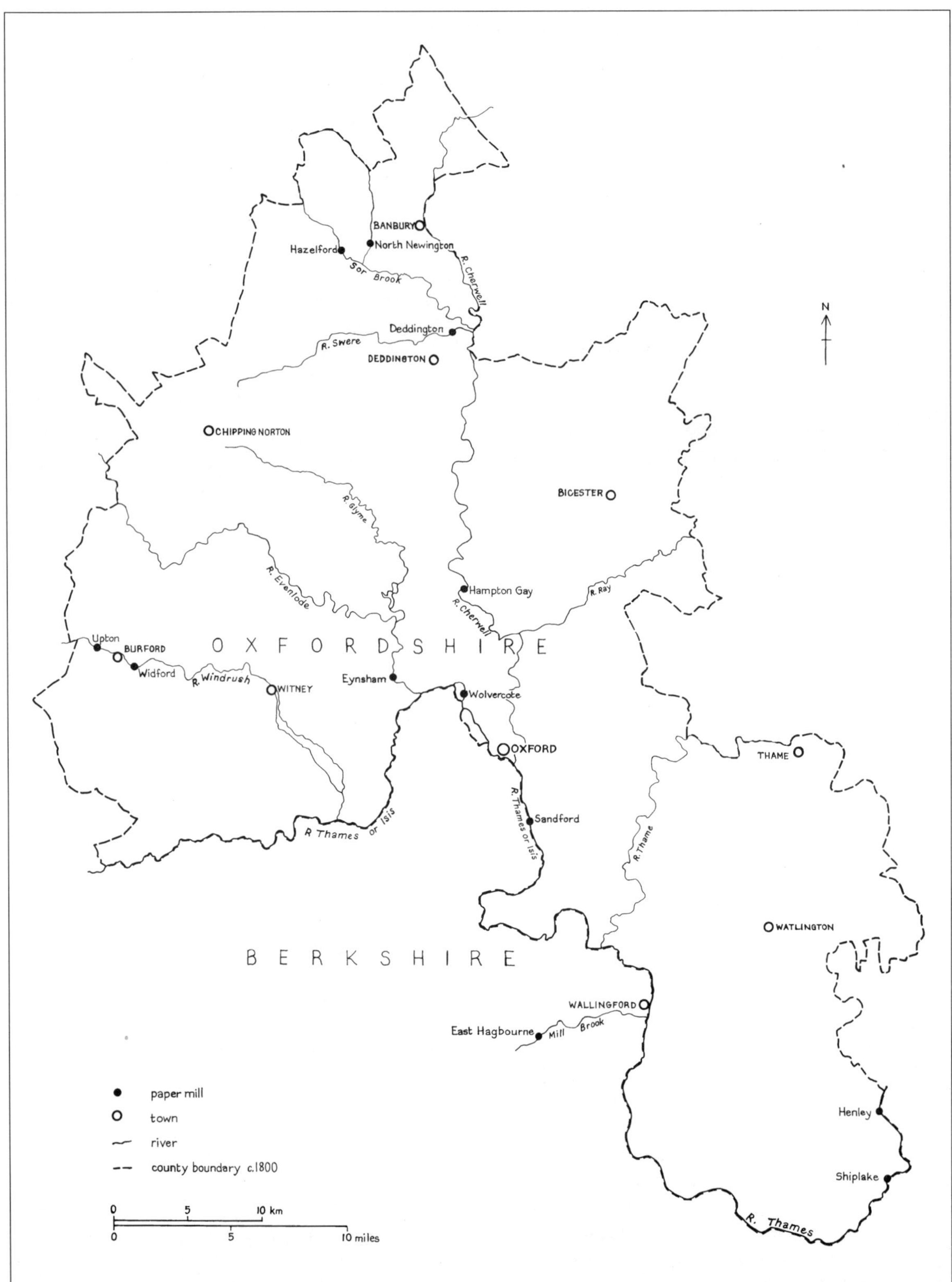

Figure 1: *Map of the Paper Mills in the Oxford Region,* (Reproduced from O.S. mapping with the permission of the Controller, H.M.S.O.© Crown Copyright).

Figure 2: *Eynsham Mill, Watercolour by James Buckler, 1826.* (The Bodleian Library, Oxford. MS TOP, Oxon. a66, fol 254)

George Hagar, a London Dyer was granted a patent for an improved method of making white paper by the sizing of the pulp in the mortar and he set up a mill at Eynsham.[11] There is some confusion over the ownership of the mill as in a lawsuit of 1686 Theophilus Poynter, an Oxford surgeon, was stated to have persuaded the owner to sell the mill to John Jordan and he let the mill to Thomas Meale(s).[12] His initials appear in the watermarked paper of *The Ladies Calling* by Richard Allestree, printed in Oxford in 1693.[13] On his death in 1706 his son Thomas took over the mill and continued there until his death in 1723.

It is possible to have an idea of the size of the mill from the Inventory made when Thomas Meale II died.[14] The mill had two vats and there were six moulds and other equipment, 25 tons of rags as well as 140 reams of paper all valued at £265 0s 6d. Meale was also engaged in farming and the details are also in the Inventory. Eynsham was supplying the University Press with paper for bibles and prayer books and was making 120 reams a week.[15] During the 18th century there were several changes of papermaker and cases of absconding apprentices. By 1785 Stephen Faichen and James Swann were the papermakers and they submitted specimens and tenders for a Royal

wove 63lb and an open Demy 46lb. to the Clarendon Press. When Faichen died in 1804 Eynsham was purchased by John Swann for his brother James.[16] The Swann family were prominent in Oxford papermaking, having interests in Wolvercote and later at Sandford.

James & Elizabeth Swann installed a Fourdrinier machine in 1807 and at that there were ten vats. Some years later Swann invited Bryan Donkin to make improvements to the machine. A 60 inch machine was installed together with a coalfired boiler and steam engine. Apart from supplying white paper James Swann supplied tarred paper in 1811 for covering the roof at Tew Lodge[17]; paper that was also used for the roof of the mill and the dwelling house. Paper was still being supplied to the University Press and also to William Cobbett for his *Weekly Political Register* on extended credit. The Swanns were close friends of Cobbett and he rented a house in Eynsham and visited the mill. The details of his business association with Eynsham and much about his debts are in the Swann-Cobbett correspondence in Bodley.[18] Dibdin mentioned in *The Bibliographical Decameron...* 'that he has asked James Swann to supply paper for his 3rd and 4th volumes of *Bibliotheca Spenceriana...* and for the present work.'

Figure 3: *Sandford on Thames Paper Mill, c1870.* (The Photographic Archive, Centre for Oxfordshire Studies, Central Library, Oxford)

James Swann was less active in the mills from 1837 and his son Henry continued. In 1848 the firm was bankrupt and the assets at the three mills were sold. However the firm of Swann & Blake occupied the mill from 1854 until 1863 selling half stuff to Wolvercote mill.[19]

Thomas Routledge came to the mill in 1856 having decided that esparto grass was suitable for making pulp and wanting to make paper on a mill scale.[20] He was influenced in his choice of mill by the experience that some of the workers had in making paper from twitch grass on which the *Proceedings of the Royal Agricultural Society Meeting* at Oxford in 1832 had been printed. Routledge took out his first patent in July 1856 and a second in Feb. 1860. He wrote a pamphlet *A Few Remarks on Paper and the Application of a new Fibrous product...* printed on paper described as 'made from the new materials as Eynsham Mills' (undated).

Routledge sold his half stuff made into sheets to the trade, including Wolvercote. By 1860 he had ceased to sell half stuff and decided to move to Ford Mill, Sunderland as he had problems with the disposal of effluent into the Thames and the inland situation hindered production.[21]

From 1872 onwards the Wakefields operated the mill producing printing news and cartridge, tub and engined sized writings and in 1881 employed over 100 people. The Eynsham Paper Mill Co. was formed in 1888 with Stephen Wakefield as managing director and a prospectus was issued to raise £25,000. By 1891 two machines were producing 40 tons a week but in 1892 the company went into liquidation.[22] The mill was put up for sale and a photocopy of the sale catalogue exists. The closure of the mill was due to competition from imported paper, the mill's remoteness from good rail links and the flight of the managing director may have been an added factor. The mill buildings were demolished in the late 1920s but the Mill House dated 1691 has survived.

Sandford

The mill was situated on the Thames, three miles south of Oxford, on the site of a medieval corn mill. The site was well chosen as one of the deepest falls on the river provided the mill with an adequate flow throughout the year. In 1823 the corn mill was for sale and it was bought by James Swann. From the drawing by Buckler of 1826 the building was L shaped, the longer arm running downstream.[23] The mill had two water wheels. In 1840

Swann installed a 52 inch machine with one drying cylinder for the manufacture of brown and packing paper[24]. Swann died in 1846 and the firm went bankrupt in 1848. The sale of the mill took place in 1850, the contents including a five foot Fourdrinier by Donkin, with drying machine, glazing rolls; millhouse with two breast wheels, driving washers, beaters, steam boiler, etc.

Granvill & Pixey who made paper for the newspaper *Reynolds News* were operating the mill by 1861 but the freehold was held by J.T. Norris of Sutton Courtney mill. During his lifetime the mill was considerably enlarged and probably the chimney was built then which became such a conspicuous feature on the river. In 1872 when new machinery was being installed a fire broke out destroying the major part of the building. The loss was estimated at £30,000. The mill was rebuilt and Norris & Co. continued in business until 1880 when the University, through the Delegates of the Press, bought it as the weirs controlled the level of the river in the reaches about Oxford.[25] A Hercules water turbine was installed and Alfred Cannon was the manager.

In 1882 the mill was let to him and he was joined by George Clapperton. A board machine made by Black Clawson of America was installed in 1872 and in 1886 a steam engine as a standby for lack of water power, although water power was still being used into the 1930s. A second turbine was supplied by Messrs. Gilkes of Kendal in 1905 replacing one of the wheels; the other wheel chamber being converted to a Penstock.

One interesting feature at Sandford is the considerable effect the mill had on the growth of the village. Two years before the conversion to paper the population numbered 193 mainly engaged in agriculture. By 1831 the population had risen to 229. In 1840 Swann was employing 40 people and he built six cottages in Mill Row to house them. Another row of houses was built by a local shopkeeper and these housed many of the paper workers. After the Swann bankruptcy many houses were empty but in the 1870s the mill was employing about 90 workers, many of them women. More houses were built in the 1890s and by 1930 the mill owned one third of the village.

Cannon and Clapperton was merged into Alders of Tamworth in 1967, manufacturing buff and coloured manilla for labels, cartridge paper and ticket middles. Production was nearly 100 tons a week. Sandford finally ceased production in the 1980s and the mill and its chimney were demolished in June 1984 to make way for a residential estate.

The paper mills in the north of Oxfordshire were operated at various times by paper-making families such as the Huttons (Hattons), Emberlins and the Sellers. Most of the mills were of the pre-industrial era.

Deddington
The site is about nine miles north of Oxford on the River Swere, a tributary of the Cherwell. There were three mills in the area but only the Old Mill was converted to paper when Michael Hutton of Hampton Gay leased the mill in 1684.[26] On his death in 1716 the Inventory valued the paper, rags and other material at £29 2s 4d.[27] Paper making by the Hutton family continued until the mid century but by 1767 John Emberlin is recorded at the mill when he took an apprentice.

On his death in 1799 he left the mill to his son, John. The Emberlin family ran the mill for many years until 1854 when Mrs. Elizabeth Emberlin was the owner.

In 1870 the mill was reconverted to corn and Christ Church bought the site. According to Mr. R.V. Clarke some of the cottages survive and some papermaking equipment is preserved.

North Newington
The mill site is two and a half miles south-west of Banbury on a tributary of the River Sore. A water mill, known as 'Collesmille' was recorded here in 1444. In 1684 it was recorded as a paper mill and had possibly been converted by the Fiennes family to paper before the late sixteenth century.[28] Shakespeare may have been referring to this mill when in Henry VI Jack Cade taunts Lord Saye with building a paper mill 'contrary to the king, his crown and his dignity.'

Nathaniel Hutton rented the mill in 1687 and was followed by Michael Hutton.[29] During the next decades there were several changes of tenant but by 1792 the mill was insured by Thomas Cobb, John Pain and William Judd of Banbury and they continued in business there until at least 1802.[30] Some of their watermarked paper can be found in books in Philadelphia and Delaware.[31]

However by 1816 William Emberlin was the papermaker. The 1832 Sale notice describes the contents as 'with four engines for rags, a paper-making machine, a drying apparatus and all other requisites'.[32] The mill then passed to William Sellers and according to the 1851 census he was aged 51 and employed five men. By 1861 his widow Rebecca had taken over and was employing six men and fourteen women. She continued there until 1871, producing hand made writings and book papers, probably for the area around the market town of Banbury.[33] Her son diversified and the mill became a bone-grinding factory.

Hazelford

This mill was on a tributary of the River Sore. One of the mills on the Fiennes estate known as the Upper Fulling Mill was converted to paper by 1792.[34] Francis Tidbury, paper maker of Northampton insured the mill and leased it to John Sellers, who was followed by William Sellers in 1832 until his death in 1851, after which paper was no longer produced.[35]

Burford

Upton Mill was situated on the River Windrush about a mile west of the village church. The earliest record of papermaking is in 1687 when 'Thorne's boy was appointed to Old Quelch the paperman for £3'. Quelch died in 1696 and two years later Peter Rich bought 'The Bull' with sixty-three acres and worked the mill. He was a business man as well as a farmer and various records indicate that he was short of working capital. In *The First Minute Book of the Delegates of the Press* in 1719 mention is made of Mr. Rich supplying 400 reams for a long primer Testament 12mo. to the Press.

By 1763 Samuel Millbourn, paper maker, was advertising 'that he would be glad to serve all tradesmen, shopkeepers and others with several sorts of paper'.[36]

Various paper makers followed Millbourn and the sale notice in 1778 described the mill as having two vats.[37] By the early 19th century there were three vats and the Hatton family occupied the mill, followed by William Emberlin in 1829. He became bankrupt in 1835 and the mill was sold.[38] In the mid-century the buildings were demolished and only shallow depressions remain today.

Widford

The site is near Burford on the banks of the River Windrush. In 1755 the paper mill was insured by Thomas Hatton[39] and the Hatton family probably continued there until at least 1803 when J. Hatton was the master papermaker. The mill was worked from a large pond and it would seem that two mills were worked in tandem. The last recorded papermaker was John Hart in 1851.

References

1 See the report of a prosecution of Cannon of Sandford, *Paper Trade Review*, 17 Aug. 1883 p.3; 24 Aug. pp.17,18; and H. Carter *Wolvercote Mill...*, Oxford, 1957 pp.8,9.
2 Lease in the custody of Long Crendon Manor abstracted in Bodl. Ms. TOP OXON c328/1, No.24.
3 Shorter, A.H. *Paper Mills and Paper Makers in England, 1495-1800*, Hilversum, 1957, Appendix J p.408.
4 Land Tax Assessments. County Record Office, Oxford, Bodl. G.A. OXON b 85a(32).
5 *Report from the Select Committee on Fourdrinier's Patent*, House of Commons Paper 1837 No.351 p.37.
6 Bodl. GA OXON 85a 332.
7 Dep c125 Bodl.
8 Recorded in H.E.S. Simmon's *Data on the Watermills of England, in the archives of the Science Museum Library, London.*
9 *Jackson's Oxford Journal*, 2 Jan. 1875 and wood engravings in *The Illustrated London News 2 Jan. 1875 No.1847 vol. LXVI.*
10 *Victoria County History, Oxfordshire*, vol. vi, Oxford, 1959, p.158.
11 Historical Mss. Comm., *13th Report*, Appendix, part 5, p.496.
12 *Wise and others v. Jordan* (1686), Chancery Proc., Mitford's Div. Bundle 345, No.179, P.R.O.
13 Shorter, *op. cit.*, w/m 110 and 111.
14 Bodl. Ms. WILLS OXON 93 fol.296V and 107 fol 305, 208 fol 385.
15 Assignment of mortgage, Brooke and Baskett to Latane, 23rd May 1720, O.U.P. Printer's records.
16 Land Tax Assessments. County Record Office, Oxford.
17 Loudon, J.C. *An Account of the Paper Roofs used at Tew Lodge,* (1811), p.5.
18 Bodl. Ms. ENG. HIST. c33.
19 *Post Office Directory, Oxfordshire*, 1854; Wolvercote Mill accounts 1855-69. O.U.P. Printer's records.
20 Obituary notices of Routledge in *British & Colonial Printer and Stationer*, 29 Sept., 1887, p.10; *Printers' Register*, 6 Oct. 1887, p.73; *Sunderland Weekly Echo*, 23 Sept. 1887.
21 Evans, Joan *The Endless Web* (1955) pp.112-113.
22 *Directory of Paper Makers*, 1894.
23 Drawings by J.C. Buckler in the custody of the Printer to the University.
24 *The Paper Makers' Circular*, 10 Nov. 1896.
25 Delegates Orders, 1881-92, p.30, Clarendon Press.
26 Northants Record Office, (A) 5990, 5996.
27 Wills, Deddington 133/5/12, County Record Office, Oxford.
28 V.C.H. *op. cit.*, v.9, note 44.
29 V.C.H. *op. cit.*, note 46.
30 S.F.I.P. 597515, 5 March, 1792.
31 Thomas L. Gravell & George Miller, *A Catalogue of Foreign Watermarks Found on Paper Used in America, 1700-1835*, Garland Pub. Inc. N.Y. 1982, w/m 156, 157 & 158.
32 *London Gazette*, Dec. 1832.
33 *The Paper Makers' Directory.*
34 Bodl. Mss. d.d. OXON c7 1792.
35 *The Post Office Directory*, 1847.
36 *Jackson's Oxford Journal*, 29 Jan. 1763.
37 *J.O.J.*, 25 April 1778.
38 *Reading Mercury*, 13 April, 1835.
39 S.F.I.P. 148206, 18 August 1755.

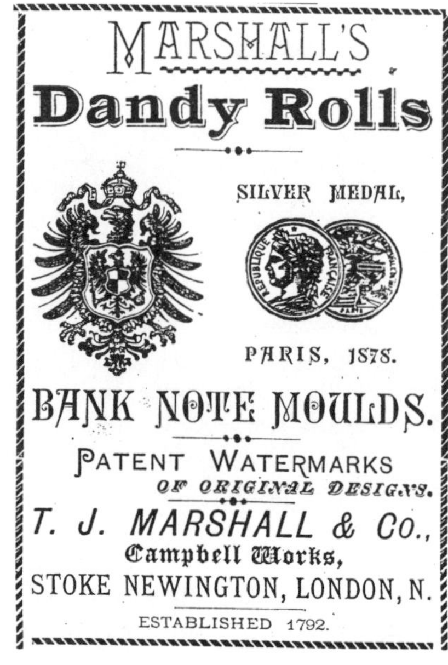

THE WOLVERCOTE MYTH

Peter Foden

Wolvercote Paper Mill's history is punctuated with superlatives. It was perhaps the first English mill to produce white paper suitable for printing - as early as 1674. It was in the first generation of English Fourdrinier machine mills, Swann brothers having had its machinery installed by Donkin in 1810. Although derelict for seven years in the mid-nineteenth century, it is still numbered among the remaining working papermills in the U.K. And finally, after complete re-building during the fifties, it became in 1965 the first fully computerized paper production line anywhere in the world.[1]

And yet, if BAPH conference delegates were to ask workers at the mill or Wolvercote residents upon what the historical importance of Wolvercote Mill depends, their reply would include none of these significant facts. They would answer in just three words: *Oxford India Paper*. If you probed further, you would be informed that this product, famous for its thinness, whiteness, and opacity, was invented and manufactured for the Oxford University Press exclusively at Wolvercote.

The tradition that India Paper was invented and manufactured at Wolvercote can be traced back to 1896. An article was then published in the *Publishers' Circular* and *The Periodical* (the free publicity magazine of the Oxford University Press), stating that "the incidents which led up to the manufacture of the Oxford India Paper have never yet been told." The story unfolded:

> In the year 1841 an Oxford graduate is said to have brought home from the Far East a small fold of extremely thin paper, which was manifestly more opaque and tough for

Figure 1: *The Paper Mill at Wolvercote, Oxfordshire, belonging to James Swann Esquire, watercolour by James Buckler, 1826.* (Oxford University Press, Thomas Photos 66223).

its substance than any paper then manufactured in Europe. He presented it to the University Press. The late Mr. Thomas Combe ....Printer to the University, found it to be sufficient for twenty-four copies of the smallest Bible then in existence - Diamond 24mo - and printed an edition of that number, which bore the date 1842. ....All Mr. Combe's efforts to trace the paper to its source were futile, and, as the years rolled on, the circumstance was forgotten. But early in 1874 a copy fell into the hands of Mr. Arthur E. Miles, of the firm of Messrs. Hamilton, Adams & Co., who showed it to Mr. Frowde [Publisher to the University], and experiments were at once set on foot at the Oxford University Paper Mills at Wolvercote with the object of producing similar paper. The first attempts were failures, but, before long, success was achieved, and on August 24, 1875, an edition of the Diamond 24mo Bible, similar in all respects to the twenty-four copies printed in 1842, was placed on sale by the Oxford University Press and Messrs. Hamilton, Adams, & Co. ....The secret of its manufacture is known to only three living beings.[2]

The same story was told in the catalogue of the Oxford University Press Paper Exhibit at the Paris Exhibition of 1900, which included (as if to emphasize the Wolvercote connection) photographs of the Mill and notes about its history.

Let us test the veracity of this piece of publisher's publicity, at least as far as we can test it against the extant written primary sources in the archives of the University Press. Firstly, we have the story of the Oxford graduate bringing back a sample of paper from China or Japan. Secondly, the printing of a very limited edition of a small Oxford Bible on that sample of paper. Thirdly, after the rediscovery of that book in 1874, experiments carried out at Wolvercote Mill to replicate such paper, and finally the successful production of Oxford India Paper at Wolvercote and the publication of a wide range of Oxford Books on this superior product.

Against the tale of the "Oxford graduate" we may bring the knowledge that experiments to emulate oriental papers were not new in 1842. Nor was Thomas Combe primarily responsible for the printing of Bibles in 1842, although he had become a partner in the Bible Press in 1841. He was at that time the Superintendent of the Learned Press. He seems to have been brought into the

Figure 2: *Henry Frowde, the "inventor" of Oxford India Paper.* (Oxford University Press, Thomas Photos 65961).

Figure 3: *The Oxford India Paper Trade Mark.* (Oxford University Press).

story to bolster the Wolvercote connection, since from 1855 until 1872 he was nominally the owner of the Mill which he had rebuilt in 1855 and made the sole supplier of the University Press with paper. The *Combe memorandum*, a note on the fly-leaf of an experimental thin paper Bible in his handwriting and dated 1853, does indeed prove that he was searching for a thin printing paper, but this was two years before he bought Wolvercote Mill, so does not indicate experimentation there. He also in fact instituted enquiries at a Fourdrinier Mill in Staffordshire, which manufactured *Pottery Tissue*, as a possible source of thin printing papers.[3]

A lightweight Diamond 24mo Bible was indeed printed at the University Press in 1842 however, and the copy acquired by Henry Frowde in 1874 survives among the Press archives. The paper used is off-white, moderately opaque, and has a weight of approximately 28gsm. Its texture is remarkably even, with no holes or paper-makers' tears, and the smooth surface suggests that the sheets were hot pressed after printing (this was routine with the later Oxford India Papers).[4] It could well be oriental. The "small fold" of the story would however have had to have been a well-wrapped parcel of about 4kg weight! The "prototype" Oxford India Paper therefore existed and has survived for paper historians to scrutinize.[5]

It becomes more difficult to establish the facts about the "invention" of Oxford India Paper in 1874-5. What was the product upon which Frowde's Diamond 24mo Bible was printed? Frowde must have had many discussions with the printers at Oxford and manager of the mill at Wolvercote during 1874, but was careful to commit nothing to paper as to the search for a suitable light-weight paper.[6] Edward Pickard Hall, the managing partner in the University Press at Oxford, added a postscript to a letter dated 6 October 1874, "The Diamond 24° on Chinese paper shall be well considered". Nothing further is mentioned in his official letters either. Frowde was also at the time in correspondence with John H. Stacey, another of the partners, who had previously been the manager of Wolvercote Mill under Combe, but nothing is mentioned about the search for India Paper. By 10 November 1874 suitable paper had been found.[7] But where?

At this point I must reveal some evidence out of sequence. Between 1888 and 1964, Oxford India Paper was made for the Oxford University Press at a paper mill in Staffordshire belonging to Brittains Limited. The earliest extant contract between OUP and Brittains contains the following preamble:

Figure 4: *Joseph Castle, Controller of the Oxford University Paper Mill, From the Supplement to the Paper Maker and British Paper Trade Journal, July 2nd 1906.* (Oxford University Press).

Whereas the Firm of Thomas Brittain and Son ...has for some years past manufactured Pottery and other thin Copying and trans-ferring Tissue papers ...and have also manufactured for the exclusive use of the University Press certain special thin opaque printing papers in accordance with and by means of special processes and methods of manufacture *known only to the said Firm or some of its Employees* in pursuance of an arrangement or understanding between the said Firm and the University Press that the said Special Printing Papers so manufac-tured as aforesaid and no other papers should be used by the University Press for the purpose of the various Oxford Editions of "Thin India Paper Books" published by the same Press.[8]

Figure 5: *Wolvercote Mill at the turn of the century.* (Oxford University Press, Thomas Photos 66224).

This agreement goes on to allow for the future additional manufacture of India Paper at Wolvercote Mill. The implication is however strong that Brittains are the primary manufacturers and possess the secret. They must therefore have been the inventors of "Oxford India Paper", which must share some of the characteristics of their "Pottery Tissues", used for transferring designs to pottery. Combe's findings of 1853 had finally been put to use.

If we look to the ledgers of the University Press, we find that the first deliveries of thin paper for the Diamond 24mo Bible in May 1875 did indeed come from Brttains, and not from Wolvercote.[9]

Wolvercote however appears to have gone into production during September 1875, and continued to supply large quantities of thin paper to the University Press until April 1876. Thereafter relatively small and sporadic consignments from Wolvercote are indicated in the Paper Journal.[10] Meanwhile the purchase ledger of the University Press shows regular large payments to Brittains for *thin paper* commencing at 7 July 1876.[11] Are we then dealing with two different products, one developed at Wolvercote, and the other in Staffordshire, or had Brittains divulged their secret to the Manager of Wolvercote Mill? Secondary evidence suggests the latter.[12] In either case, we are left to assume that the Wolvercote product was inferior in quality, and that production ceased altogether after April 1877 - a significant admission of failure, since the policy of the Press since 1855 had been to use only Wolvercote papers. An informal agreement may have been made between the University Press and Brittains before October 1877, since from that date a discount of 2.5% was applied to deliveries of *thin paper* from Brittains.[13] For almost a century, Brittains remained the exclusive supplier of Oxford India Paper. Successive agreements allowed for manufacture at Wolvercote (to Brittains' furnish) only when the Cheddleton Mill failed to meet all Oxford's demand. A copy of the furnish was therefore made by Joseph Castle, the Controller of Wolvercote Mill, in 1888, and records makings of India Paper there during 1888-90, 1904, 1909, and 1912. The raw material was "solid tarred hemp hawser coil", but the secret lay in a repeated process of boiling, pressing, willowing, and dusting, lasting in total well over 24 hours.[14]

Frederick Haigh of Brittains believed that Oxford India Paper as manufactured from 1875 was not so much a new invention as a revival of an old one. In a letter to Horace Hart dated 30 October 1896, he claimed that the 1842 Diamond 24mo Bible was printed, not on oriental paper, but on "Pottery Tissue" made in Hanley by Fourdriniers (Brittains' predecessors). Brittains possessed no archives, but Haigh wrote of "a vague tradition down here in his early days that a paper of this kind had been made altho it was not known who for".[15] A similar statement was made by Harold Hart in 1930.[16] Such an origin of the prototype for Oxford India Paper is not borne out in the archives of the Press: the Bible Press had three suppliers of paper in 1842, and Fourdrinier was not one of them. If the paper were made in this country by machine, then Dickinsons would be the likeliest source.[17]

Henry Frowde was well aware of the true nature and origin of Oxford India Paper, but chose to use the more romantic story of the traveller in the Far East and experiments carried out under the legendary figure of Thomas Combe at Wolvercote Mill for advertising purposes. His early success as University Publisher depended in part on the "discovery" of this new product, which he marketed with all the wily skill of a nineteenth-century Bible publisher.[18] The Wolvercote connection was also considered advantageous for the registration of the trademark Oxford India Paper when faced with rival thin printing papers in 1912.[19] Harry Carter's definitive history of Wolvercote Mill set matters straight in 1957, but the Wolvercote Myth, an advertiser's dream,[20] remains firmly embedded in local tradition.

Acknowledgements

I would like to thank the Secretary to the Delegates of the Press for permission to publish this article based largely upon the archives of the Oxford University Press, and expert advice given me by Richard Neave about samples of Oxford India Paper.

References

1   Harry Carter *Wolvercote Mill: A Study in Paper-making at Oxford* (Clarendon Press, 1957); *The Wolvercote Story* (1965).
2   *The Periodical*, December 1896, p.18.
3   Cox and Timberlake's Notebook [OUP Archives] p.26; discussed in Carter, *Wolvercote Mill: A Study in Paper-making at Oxford* (Clarendon Press, 1957), p.43.
4   I am grateful to Richard Neave, OUP's head paper buyer, for making these preliminary observations about the paper.
5   This volume was believed lost when Harry Carter wrote his history of Wolvercote Mill in the 1950s.
6   Henry Frowde's copy letterbooks survive for his entire career as Publisher to the University (1874-1913).
7   Frowde to Hamilton, Adams, & Co., 10 November 1874 [Frowde letterbook 2, folio 92].
8   Agreement between The Delegates of the Clarendon Press in the University of Oxford and Thomas Brittain, Thomas Arthur Brittain, Joseph Cecil Clay, and Frederick Haigh, 20 February 1888.
9   Paper Journal 1854-81, May 1875.
10 Reamage of thin Bible paper supplied to the University Press by Wolvercote Mill (whole reams only):

1875	September	2
	October	150
	November	203
	December	209
1876	January	114
	February	331
	March	491
	April	329
	May	3
	June	5
1877	March	5
	April	22

11 University Press Ledger No. 1 [1869-81] folio 348.
12 Frederick Haigh to Horace Hart, 30 October 1896.
13 University Press Ledger No.1 folio 348; the 1888 agreement between OUP and Brittains implies a pre-existent understanding.
14 Notebook labelled "India Paper" in Joseph Castle's hand. Contained in sealed envelope which was opened at the Conference. Previous openings had taken place in 1920 and 1956.
15 Frederick Haigh to Horace Hart, 30 October 1896.
16 Harold Hart, Brittains Limited, to John Johnson, OUP, 2 October 1930 (Harold Hart was Horace Hart's son).
17 Annual Summaries of Account of the Bible Press Concern, 1841, 1842, 1843: the Press was supplied with paper by Dickinson, Swann, and Venables.
18 Gross Sales by the Bound Book Business of the University Press grew from £22,000 (1874) to £40,000 (1875) and £53,000 (1876): Frowde took charge on 25 March 1874 [Private Business Record, OUP London].
19 Case for the Opinion of Mr. Sebastian, Rivington & Son, 1912: "In or prior to 1875 The Delegates of the Oxford University Press after numerous attempts succeeded in manufacturing at their Wolvercote Works ....India Paper...."
20 The "advertiser" responsible was probably Robert Maynard Leonard, the first Publicity Officer employed by the Oxford University Press: it is probably more than coincidence that his full time employment began in 1896.

BENTLEY AND JACKSON,

Engineers, Ironfounders, &c.,

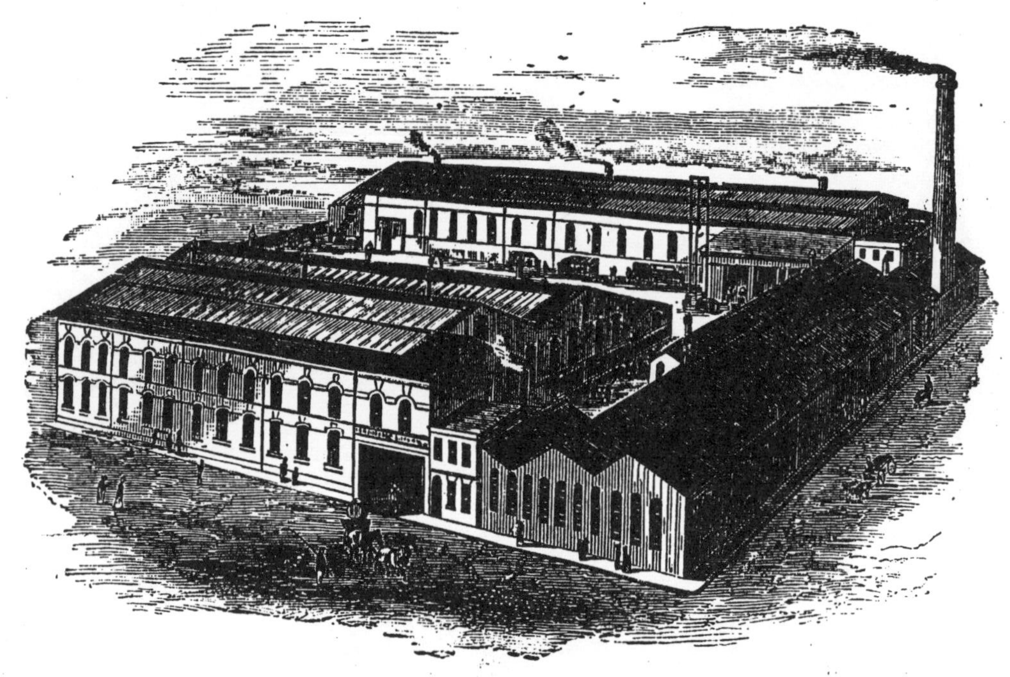

LODGE BANK WORKS,

BURY, NEAR MANCHESTER.

MAKERS OF ALL KINDS OF

Paper Mill Machinery,

CHILLED IRON ROLLERS,

Hydraulic Pumps and Presses.

[ESTABLISHED OVER TWENTY YEARS.]

STRAW

James Brander

Introduction

Cereal straws (mainly wheat and rice) have been used as a source of papermaking fibre for at least 300 years and now provide about 4% of the world's total pulp supply. Over this period the chemistry and much of the mechanics of the processes used to make this pulp have changed little. Patents filed in the 1850s describe the technologies used today for the major part of the straw pulp made: a simple soda or lime cooking liquor, temperatures up to 170°C, some mechanical action derived from the tumbling of a rotary digester vessel and cooking time of several hours describes how perhaps 80% of the world's straw pulp is made. Mills using technology of this sort are small, no more than 40 tonnes per day, and supply slow paper machines.

The furnishes used on these machines are usually not pure straw, but blends with waste fibre, bamboo or wood pulp. The products made cover the full range of papers and boards needed in society.

Especially during this century, a wide variety of different processes has been tried in straw pulping, achieving popularity for a few years, but rarely being replaced by a second generation. Examples are: the Pomilio process: screw-fed, horizontal tube pulpers of the Pandia type: the NACO process: anthraquinone.

The enthusiastic investment in horizontal tube pulpers which occurred in the '60s and '70s allowed the development of some relatively large mills, producing up to 100,000 t/year. Some half dozen of these are in operation today, though they must be getting a bit antique. One has just shut, possibly temporarily. Despite its fairly long history as a pulping material, and its undoubted good points as a papermaking fibre, straw has probably never been more than a substitute for preferred, but less available fibres. In the seventeenth and eighteenth centuries the preferred fibre was rags: then it was esparto; now it is wood. Straw is now used only in countries whose access to international commodity markets is, or was until recently, restricted.

There are many reasons for this lack of enthusiasm for straw as papermaking pulp, but they come down to two most important: its seasonality and consequent difficulty of storage: its slow drainage rate. These characteristics have direct consequences for the costs of producing paper from straw.

However, there are estimated to be in excess of 1000 million tonnes annually of cereal straw available around the world, which theoretically would satisfy the world's pulp demands, even at a yield of only 20%. So there ought to be a lot of interest in trying to utilise straw more fully.

This paper will look at some of the shortcomings of straw as a pulp material and try to describe what work is in progress to circumvent these.

History

Most of this chapter is taken from Richard Hill's book.[1] Several patents on papermaking fibres were granted during the seventeenth and eighteenth centuries as manufacturers tried to diversify out of rags. Straw wasn't explicitly mentioned in any of them, though it is thought that it was widely used as a filler pulp in at least the eighteenth century. It was not generally considered very good, though even Turner used it when necessary.[2]

The first explicit reference to straw pulp is in 1800 when Matthias Koops used straw paper for the first edition of his history of communications. A year later however, the second edition came out on recycled paper. Koops did take out a patent on straw pulp production in 1802, in which he described having to cut the stalks into two inch lengths, a problem which has recently attracted several hundred thousand ecu of EC Framework Programme funding.

An 1826 patent described alkali cooking, chlorine bleaching and the problem of silica, which the EC hasn't dared tackle yet. Interest in straw seems then to have waned rather until the 1850s, when a patent describing high temperature and pressure cooking with alkali was published. Straw was in regular use around that time, especially in Europe, but was regarded as expensive and wasteful, and as producing inferior quality paper. Straw then came into its own again during the World Wars, especially the second.

Thus, it seems that, in Britain, interest in straw resurged: during the Napoleonic wars, when rag and paper supplies from France were cut: during the American civil war when cotton supplies from the Confederacy states were cut: moderately during the first world war when shipping became expensive; and during the second world war when the North African esparto supplies were cut. In other words, straw was only used here when preferred material was unavailable.

Current straw pulp production around the world
According to the FAO[3], some 10 million tonnes of straw pulp are made in 23 countries around the world. Those producing more than 10,000 tonnes per year are listed in Table 1 below.

China, India and Pakistan account for 9 million tonnes, while 8 European countries contribute a further 360 thousand tonnes (at least, they will when Denmark's Fredericia and Hungary's Duna mills restart). Wheat straw is more commonly used than rice, despite the main producing areas being substantial rice growers too.

Only China is increasing its straw pulp production, though Turkey and India keep talking about it. (I believe the latest Indian pulp project I saw described was based on wood.)

It is probable that most of the digesters in use are more than thirty years old, though some of the larger mills have invested in new peripheral equipment in the past fifteen years, mainly for environmental reasons.

Sizes of straw pulp mills
Straw mills are mostly tiny by modern kraft pulp mill standards, typically producing between 10,000 and 40,000 tonnes per year of pulp. China's relatively substantial pulp production is all from mills of that sort of size: there were more than 1,000 of them in the early 1980s. Usually the straw is supplemented by another fibre source (often waste paper).

Some of the smallest straw pulp mills are in Italy, with production capacities of under 1,000 tonnes per year. There are some larger mills. The now-closed (temporarily?) Duna mill in Hungary had a total output of 300,000 tonnes/year, though by no means all of this was straw. Calarasi in Romania was started with 60,000 tonnes per year production capacity, catered for with two screw-fed continuous digesters. Its capacity now is around 80,000 tonnes per year, with the straw pulp supplemented by waste paper. The Foggia mill of the Italian graphics research institute has a capacity of 75,000 tonnes/year, though this too is not all straw.

The characteristics of straw and its fibres
Cereal straws are of course agricultural waste. Structurally, they consist of leaves, nodes and internodes. The internode theoretically has both bast and pith, of which the latter is usually missing. Table 2 summarises the proportions of some of the interesting constituents of each part in the case of barley straw.[4]
Table 2 suggests that it would be beneficial in terms of pulping yield, protein content, and inorganic material to separate the leaf from the internode before pulping, and

TABLE 1 - MAIN STRAW PULP PRODUCING COUNTRIES[3]

Country	Metric Tonnes per year
China	8,400,000
India	400,000
Pakistan	142,000
Spain	142,000
Turkey	77,000
Egypt	67,000
Romania	58,000
(Denmark)	51,000
Indonesia	42,000
Syria	32,000
Hungary	30,000
Italy	30,000
Iran	30,000
(Yugoslavia)	28,000
Algeria	25,000
Greece	16,000
Bulgaria	12,000

TABLE 2 - THE DISTRIBUTION OF INTERESTING CONSTITUENTS IN BARLEY STRAW[4]			
on oven-dry whole-plant basis	Leaf	Internode	Node
Weight proportion	38.7%	55.0%	4.2%
Cellulose	11.3%	23.2%	
Lignin	6.0%	9.9%	
Minerals	3.2%	2.7%	
Silicon	0.9%	0.6%	
Protein	2.0%	1.7%	

use it as separate fodder.

Additionally, the leaf and node material also includes a higher proportion of parenchyma cells than the internode. These are much smaller than the fibres, and it is their presence that so profoundly affects the drainage of straw pulps.

Barley straw, even the internode material, contains substantially more of these short parenchyma cells than wheat straw, as Table 3 illustrates.

Hence, barley straws are significantly harder to drain even than wheat straws. Rice straw is comparable in this respect to barley. The contents of both lignin and cellulose in straw are rather low. The low lignin is an advantage since it facilitates the cooking process, allowing the use of gentler alkalis than caustic soda (e.g. lime and sodium carbonate) in many applications.

On the other hand, the low cellulose content is one factor dictating why straw pulps tend to be produced with low yield.

Apart from the parenchyma cells referred to above, the most problematic constituent of straw is the minerals, particularly the silica. The example given in Table 2 isn't too bad; as shown in Table 4, the silica content can sometimes be as high as 18% in rice straws.

Table 4 shows that, except for the average length (which disguises the presence of the very short parenchyma cells referred to above), the straws are really quite different from modern preferred papermaking fibres. The fibres are very much more slender than those of wood but in fact (not shown in the table) rather thicker walled. On balance they should be less rigid than wood fibres, though they will probably tend to collapse less than softwood fibres do. There will obviously tend to be many more of these short, slender fibres per unit weight of

TABLE 3 - THE RELATIVE FREQUENCY OF OCCURRENCE OF FIBRE LENGTH IN CEREAL STRAWS[5]		
	Length	
	0-0.5 mm	1.0-1.5 mm
Wheat	0.18	0.48
Oats	0.30	0.29
Barley	0.34	0.37

TABLE 4 - WHOLE PLANT CHARACTERISTICS OF SOME PAPERMAKING FIBRES[6,7]						
on oven-dry whole-plant basis						
Material	Ave. fibre length (mm)	diameter (μ)	Silica %	Cellulose %	Lignin %	Pentosan %
Straw						
Wheat	1.5	13	4-7	33-40	15-19	22-30
Rice	1.5	10	12-18	35-45	11-14	21-26
Softwoods	2.0	30		35-45	25-28	6-8
Hardwoods	1.0	20	< 0.5	40-55	20-25	15-22

pulp than is the case with wood, though they will tend to pack down well in the paper sheet.

The chemical constituents indicate, from the lignin content, rather low chemical demand: from the cellulose content, rather low pulping yield; and from the pentosan content, also rather a low pulping yield, with possibly some interference with papermaking chemicals. These differences are summarised in Table 5.

Characteristics of papers made with straw pulp
The outstanding feature of straw-containing paper is its

fibres per unit weight of pulp. The smoothness and formation of straw-containing fibres also benefit greatly from the large number of fibres per unit weight. All these characteristics (stiffness, opacity, formation, smoothness) are highly prized in printing and writing papers.

Another feature of straw-containing papers that is attractive in printing and writing grades is the self-sizing. One of the chemical components, perhaps a fatty acid, or maybe residual lignin, actually imparts quite a high degree of water-repellency. Also, straw pulp, because of

<table>
<tr><td colspan="2">TABLE 5 - THE EFFECT OF THE DIFFERENCES BETWEEN STRAW AND WOOD FIBRES</td></tr>
<tr><td>Thinner</td><td>Better bonding
Better formation
Smoother sheet
Weaker fibre
Less bulky sheet*
More opaque sheet</td></tr>
<tr><td>Thicker walls</td><td>More opaque sheet</td></tr>
<tr><td>Silica content</td><td>Better stiffness: 'crackle'
Inefficient chemical recovery</td></tr>
<tr><td>Lignin content</td><td>Reduced cooking chemical demand
Simpler cooking systems</td></tr>
<tr><td>Cellulose content</td><td>Lower yield
Easier refining?</td></tr>
<tr><td>Pentosan content</td><td>Lower yield
Papermaking chemical interference</td></tr>
<tr><td>Parenchyma cells</td><td>Smoother sheet
Poor paper machine runnability
Higher washing costs
Higher bleaching costs
Less bleaching flexibility</td></tr>
<tr><td colspan="2">* compensated for by better sheet smoothness</td></tr>
</table>

crackle: it always feels rigid. This is probably due to the presence of silicates on the fibres, reprecipitated towards the end of the pulping process. The resistance to collapse of straw fibres means that they will have good opacification: enhanced further by the large number of

its poor drainage, requires very little refining, though this means it develops little strength. This doesn't matter in printings and writings, where advantage can be taken of this to reduce energy costs. The slender fibres should mean that straw-containing papers tend to be of high density, which is not of benefit in printings and writings, but in practice, the low refining requirement and reduced need for calendering all mean that sheet thickness is maintained.

Thus, straw really should be very much appreciated in printing and writing grades, and it is interesting that a lot of research effort is expended on trying to make wood fibres behave more like straw in these applications.

Straw's stiffness makes it very attractive too for corrugating medium, though its poor strength development renders it less suited to the outer layers of corrugated laminates or to other high-quality packaging. Similarly, straw is not suited to coating grades where tear and tensile strength are important. But, certainly, in uncoated woodfree printings and writings, straw ought to be used much more than it is.

Principal paper and board grades where straw is used
From the above discussion of its papermaking characteristics, it is clear that straw ought to be most useful in uncoated printings and writings and in corrugating medium. Certainly, in China it is widely used in these papers, and in most others. In other countries, however, it tends to be used only in the lowest quality packaging and board, including 'fibre board' for building material.

Until recently, there were reasonable amounts of market bleached straw pulp available from Denmark and Hungary, and a straw-containing paper using the Hungarian pulp was launched on the U.K. market earlier this year. Despite the proceedings of the PIRA straw symposium's being printed on it, this paper doesn't seem to have been a great success.

The reasons for this lack of enthusiasm, despite straw's many advantages, may become clearer later in this paper, where I show how straw pulp is really quite a lot harder and more expensive to make than wood pulp, though I suspect that doesn't provide the whole answer. Perhaps it is like bicycling: perceived to be low-grade even though actually highly practical in the right circumstances.

Technologies available for pulping straw
The technologies available for straw pulping nowadays fall into two types. A third, the Pomilio, is mentioned but is no longer in operation anywhere. Both types can be used with any level of chemicals, to produce fully-bleachable or semichemical pulps. Both can be enhanced with oxygen, anthraquinone or sodium carbonate.

a) *Batch tumbling digester*
The simplest and most widely used straw pulping

technology is a tumbling digester vessel, spherical or cylindrical loaded with five tonnes of straw, plus the caustic soda or lime used for cooking. The cooking vessel will be steam heated to a temperature of 170°C (maximum, often less), and cooking may continue for as long as four or five hours, depending on the initial condition of the straw and the grade of paper for which the pulp is intended. The total cycle time (loading-heating-cooking-unloading) of this process can be as long as eight hours, at a pulp yield often as low as 35%, so a mill producing forty tonnes per day will need several such digester vessels.

Various chemicals can be used in this type of vessel. Caustic soda is the simplest and used for pulp intended for bleaching. Lime gives a less harsh cook with a better yield of unbleachable pulp for corrugating medium. For a bleachable pulp, caustic would normally be added at a rate of about 15-20% on dry straw. This can be reduced if any of the additives is in use.

Ammonia can be used instead of caustic, though this gives a less bleachable pulp. It has the advantage that the effluent can be put directly onto fields, though the operation must be fairly unattractive.

b) *Screw-fed, horizontal tube continuous digester*
The screw-fed, horizontal tube digesters (Pandia) developed during the 1950s and utilised in most of the larger mills built during the '60s and '70s use basically the same cooking chemistry, but with very much shorter cooking times (fifteen to twenty minutes being typical). This, coupled with the continuous nature of the process, gives rise to much improved product quality and consistency, and better capital utilisation.

The differences between the systems are as much in the care of the straw preparation as in the actual pulping technology, though the reduced impregnation time of the screw-fed systems is a considerable factor in reducing the cooking time.

Yields are rather higher in continuous systems than in batch systems, though still not above 40% for a bleachable pulp, and chemical charges are similar too.

A variant of the horizontal tube digester was the Taystile process, in which the place of the screw feeder was taken by a large peristaltic pump, and the chemicals were added to the straw in a conventional papermaking pulper. However, this variant did not seem able to achieve the short cooking times of the screw-fed process, possibly because the impregnation was not so effective.

c) *The Pomilio process*
This is really a historical curiosity now, as none is

believed still to be operating. It was based upon the old Cross and Bevan technique for analysing cellulose content by extraction with chlorite. The process was important because it combined mechanical defibration with the chemical treatment. I understand the process to have been invented in Argentina during the first world war (presumably the Argentine was short of esparto too), though it achieved most of its application in Italy during the 1930s. It was highly polluting as the chlorinated effluent was no use for anything.

The trouble with straw

Two of the main reasons for straw's unpopularity as a pulping material have already been referred to: storage and drainage. These issues, and others, will be discussed here.

a) Detrimental effect of moisture

Straws have the intrinsic awkwardness of being harvested at most twice per year, and usually only once. This obviously leads to an enormous storage problem for any mill; one which increases drastically with size. Even with good storage, up to 30% of the raw straw can be lost in transport, washing and chopping before the digester, so that a mill producing 300,000 tonnes per year of chemical straw pulp would probably need to store 1m tonnes of straw, somehow. This storage needs to be good, since bacteria very quickly attack moist straw, reducing its cellulose content and darkening it. Ideally, the moisture content should be kept below 15%. Above 20% the straw becomes almost unusable. The driest straw from a U.K. field has about 12% moisture. The highest bale densities normally achieved are around 200 kg/tonne, which is not enough to keep moisture out, so in any temperate climate there is no escape from building roofed storage, with its attendant costs and fire risks.

b) Density

Straw has a very open structure, often cited as one reason for its relative ease of cooking. While this is an advantage in respect of cooking liquor penetration, it does imply a low density. This affects transport costs, and also the maximum throughput of a given size digester vessel. Realistically, digester vessels for straw are limited to about 100,000 tonnes per year output, while those for wood now produce 400,000 plus. Clearly this militates against large mills.

c) Brittleness

Straw is a brittle material, easily generating large quantities of fine dust during any mechanical operation. Apart from the yield loss implied, a lot of the dust simply blows away, this leads to a wide variation in the size of pieces going into the digester: hence to the impregnation times; and hence to the pulp properties. To avoid this, larger mills using continuous pulpers use wet washing stages before the digester feeders, adding to water consumption, effluent problems and yield losses. The tub grinders originally widely used to cut the straw into short lengths suitable for continuous digestion were particularly prone to this dusting, and it is significant that there has been successful research recently on less dusting methods of straw cutting.[8]

d) Poor drainage

As noted above, the main reason for the poor drainage of straws is the high proportion of small parenchyma cells throughout the plant, though fines generated by

TABLE 6 - EFFECTS OF THE MAIN PROBLEMS WITH STRAW PULPING

Detrimental effects of moisture	Costly storage
Low plant and bale density	Costly transport Size of digester vessels
Brittleness	Yield loss Increased chemical demand Variability of pulp
Drainage	Capital cost Bleaching chemical demand Paper machine runnability
Silica	Reduced black liquor solids Increased maintenance Silication of recovered chemicals

mechanical action also contribute to this problem. This issue has some serious consequences.

In pulping, it severely limits the effectiveness of washing stages, demanding the use of high water volumes and more washing stages. This increases capital and operating costs, while reducing the solids content of the wash liquor, rendering it more difficult to burn in recovery. Poor washing after pulping also increases the demand for bleaching chemicals, and may limit the choice of bleaching systems. Modern wood mills producing totally chlorine free pulps rely substantially on good washing to make their peroxide and ozone treatments effective.

On the paper machine, the poor drainage of straw is said to mean that it can't be used as the major part of a furnish at speeds above about 300 metres per minute, since it doesn't drain well enough to develop strength. However, experience with modern twin-wire machines with closed press sections is that they can run pretty well anything, though it's probable that no-one has tried straw.

e) *Silica*

Silica contents of up to 18% are said to be found in some rice straws, though in wheat the level is much lower. Still, it presents serious problems for mills trying to utilise chemical recovery systems. High silica content in the cooking black liquor raises its intrinsic viscosity at a given solids content. To circumvent this, the liquor must be kept hot and very alkaline. This of course raises operating costs. Even then, its final total solids after evaporation cannot be raised above at best 65%, so that its calorific value is reduced compared with wood liquor (usually at 75% now). Hence the thermal effectiveness of recovery is reduced.

Additionally, the silica precipitates out, especially in evaporator sections, forming thin, transparent films that wreak havoc with thermal efficiencies. It can also build up in recovery boilers, increasing explosion risks. Hence traditional recovery systems are less effective and more costly to operate and maintain than in the case of wood pulp mills.

These major problems and their effects are summarised in Table 6. They all translate into increased costs for using straw pulp rather than wood, even though wood itself is many times more expensive. The position was the same in the nineteenth century too, when straw pulp cost the same to produce as rag pulp, even though straw cost one eighth as much as rags.

Recovery

This issue has been raised in connection with silica, which was shown to render conventional recovery problematic and hence expensive. However, these systems are used in all the larger mills. The problem arises with smaller mills, for which conventional recovery is not practicable. It has been estimated that only about one third of the chemicals available for recovery from Chinese straw mills actually are recovered. The rest goes down the drain, with depressing environmental consequences.

All straw mill recovery systems would benefit from the removal of silica from the black liquor, and a lot of research has gone into finding ways of doing this. A scientifically appealing way of doing this is to lower the pH by carbonation with flue gas until the silicates precipitate out, leaving the organics still in suspension to be burned in the recovery boiler. There may be some functioning installations based on this principle; certainly attempts have been made to apply the idea.

Another scientifically ingenious idea was developed in China. Known as *Wet Cracking*, the idea this time was to heat the black liquor to 360°C, at which point the organics crack apart, yielding quantities of burnable gases and carbon dioxide. The carbon dioxide acts to precipitate the silicates, the gases are used to heat the black liquor, and the resulting liquid is a dilute solution of sodium carbonate which can be reused in the pulping liquor. Unfortunately, it doesn't seem to work.

One straw mill (in Syria) uses a direct firing system for recovery whereby the concentrated black liquor is burnt directly in a fluidised bed grate. If all goes well, the recovered chemicals are formed into pellets of sodium carbonate, which are ejected from time to time. However, there seems to be a lot of trouble with temperature variation in the burning because of variability in the straw, such that the pellets often form a solid smelt.

Probably the most promising technology for small mills, and possibly for large ones too, is the Direct Alkali Recovery System.[9] Following trials at the APPM mill in Tasmania, a commercial plant of this type was built at the Fredericia straw mill in Denmark. Sadly, it didn't work.

So, as things stand, recovery from straw mills is not fully developed, and it is probable that a small mill would just have to accept the cost of chemicals and treat its spent pulping liquor as effluent. There is possibly some escape from the worst of this by pulping in ammonia. While this isn't so effective as caustic and the pulp produced is not really bleachable, that doesn't always matter, and the effluent can be put straight onto fields as fertiliser. Research has also shown that the effluent from pulping in potassium hydroxide is particularly beneficial to various tropical crops, so this might be an option for a suitably situated small straw mill.

Current research areas

It is clear that China has the greatest interest in straw pulp, so it is fitting that quite a lot of research is being conducted by them, both at home and in association with western institutes. A large part of this Chinese-inspired research is being directed at very fundamental issues, hemicelluloses and the like, probably with a view to genetic manipulation. But high-shear, extruder-like machines operating at high consistency, and chlorine-free bleaching are also being investigated.[10]

Apart from that there is EC-funded work on milling systems to separate the leaf and node from the internode[11], and on low energy straw slicing equipment that would reduce the fines generated when straw is cut.[8] The EC also funds a programme investigating high-shear, high-consistency pulping.[12] Recent work in Pakistan has shown that even semi-chemical straw pulps can be bleached to quite high levels (ISO 80), but only using chlorine containing compounds.[13] Finally, an Indian research centre has built a pilot DARS recovery line, with which it is hoping to overcome the problems experienced with that technology.[9]

Suggestions for a straw pulping line

Consider what would be required to use straw to produce about 30,000 tonnes per year of corrugating medium. From what has been said before, the criteria should be roughly as follows:

> The mill should be integrated with a paper machine making at least 100,000 tonnes per year, so that straw constitutes less than half the furnish.
>
> The remainder of the furnish should be recycled.
>
> The pulping process should be semi-chemical, to produce unbleached corrugating medium at a pulping yield of around 60%.
>
> Wheat straw is easier to use than rice or barley: storage should be under cover.
>
> Without recovery, silica doesn't need to be removed.
>
> Without bleaching or recovery, plant parts don't need to be separated.
>
> A tumbling digester has advantages over a horizontal tube digester, but probably not enough to outweigh the short pulping time and rapid throughput for a given size.
>
> A cutter will be needed.
>
> Chemical recovery should not be attempted at first, though use of an improved DARS system may later be possible.

The equipment required would thus be something like:

> Bale opener
> Straw slicer
> Screw feeder
> Continuous tube digester
> Washers
> Effluent treatment

The cost would be of the order of £10-15m and the overall pulp yield would be around 60%, so the mill would need about 50,000 tonnes per year of straw. The production cost would probably work out at around £100 per tonne, and manning, which, at today's prices, is a lot more than waste paper would cost for the same task.

Whither straw pulp?

The example given above is not encouraging, the more so as this is the simplest system imaginable. Bleached pulp would require substantially more treatment and operate at a much lower yield. It could cost as much as £220 per tonne to produce, comparable with the present price of bleached woodpulp. And that assumes that some means of chlorine-free bleaching can be developed.

But at least it is comparable to woodpulp, whereas the corrugating medium was more expensive than its competition. And the price of woodpulp must increase soon. Government support would be needed with the capital requirement, but in a country really awash with straw, that might be forthcoming.

So maybe the future lies there, in small mills built with very long-term government loans, producing chlorine-free bleached pulp for integrated waste-based fine paper machines. Perhaps.

References

1   Hills R.L.: *Papermaking in Britain* 1488-1988, The Athlone Press, 1988.
2   Bower P. *Turner's Papers*. The Tate Gallery, 1990.
3   *Pulp and Papermaking Capacities*. 1992-97, p.119: FAO, 1993.
4   Pedersen P.B. and Munck L. 'Whole crop utilisation of barley including new potential uses' in *Barley: Chemistry and Technology* (eds. A.W. McGregor and R.S. Bhatty) Am. Soc, of Cereal Chemists, St. Paul, Minn, USA, 1993.
5   Robson D. and Hague J. 'The Properties of Straw Fibre', *Straw: a Valuable Raw Material*, v.1, PIRA, 1993.
6   Jeyasingam J.T. 'Mill Experience in the Application of non-wood Fibre for Papermaking': *1990 Pulping Conference*, p.321, TAPPI, 1990.
7   Ray, A.K., Mathur A., Varma K.V.P. and Garceau J.J. 'An attempt to Analyse Rice Straw based on Mechano-chemical Pulp for Processing': *1991 Pulping Conference*, v.1, p.223, TAPPI, 1991.
8   Knight, A. *Straw: A Valuable Raw Material*, v.1, PIRA, 1993.
9   Panda, Dr. Ing. A.: 'Recent Developments in Chemical Recovery Processes for Small Pulp and Paper Mills', IPPTA, 4(3), p.12, 1992.
10 'Recent Research in China into Straw Pulping': *Straw: a Valuable Raw Material*, v.2, PIRA, 1993.
11 Fuglsang J.: 'A Straw Milling System', *Straw: a Valuable Raw Material*, v.1, PIRA, 1993.
12 Farrar M.: 'Report on the ECLAIR Straw Pulping Project', *Straw: a Valuable Raw Material*, v.1, PIRA, 1993.
13 'Bleaching Semichemical Straw Pulp in Pakistan': *Straw: a Valuable Raw Material*, v.2, PIRA, 1993.

THE DEVELOPMENT OF THE BEATER

Phil Crockett

Introduction

This is not intended as an exhaustive or complete treatment of the subject of beaters and beating, since a vast amount has been written about the subject. I shall give a broad overview of the subject and show some details of the engineering development of this vital piece of equipment.

Fibre Treatment and Development

Even after boiling or pulping raw untreated fibres, whether from vegetable fibres, rags or wood do not bond together very well to form an acceptable sheet of paper. They all require some form of treatment.

The first stage of this treatment involves breaking the raw material into its individual fibres, this was often combined with washing and bleaching, to produce what used to be called "half-stuff".

The second stage took the "half-stuff" and developed it to suit the product being made, so we can have:

1. Cutting without wet beating - for example, blotting paper.

2. Wet beating without cutting - for example, greaseproof or tracing paper.

3. The enormous range between these extremes which cover most types of paper.

Cutting, or shortening, the fibres is self explanatory. "Wet beating" though is far from obvious and there were heated debates about it well into the 1930s as to whether it was a physical or chemical action.[1] In fact there is still a lot of debate and misunderstanding about fibre treatment, and my colleague at Arjo Wiggins R & D is continually running refining (or beating) trials for our mills, to establish the best way of treating the fibres to obtain the properties required in the paper.

The paper maker knows that his fibres have been "wet" beaten because they take longer to drain than with so-called "free" beaten paper. Wet beating involves fibrillating the fibres, or working on them to open up fibrils which increase their surface area so that they hold more water. The surface area doubles with beating from 14°SR to 28°SR, quadruples by 42°SR, and has increased by more

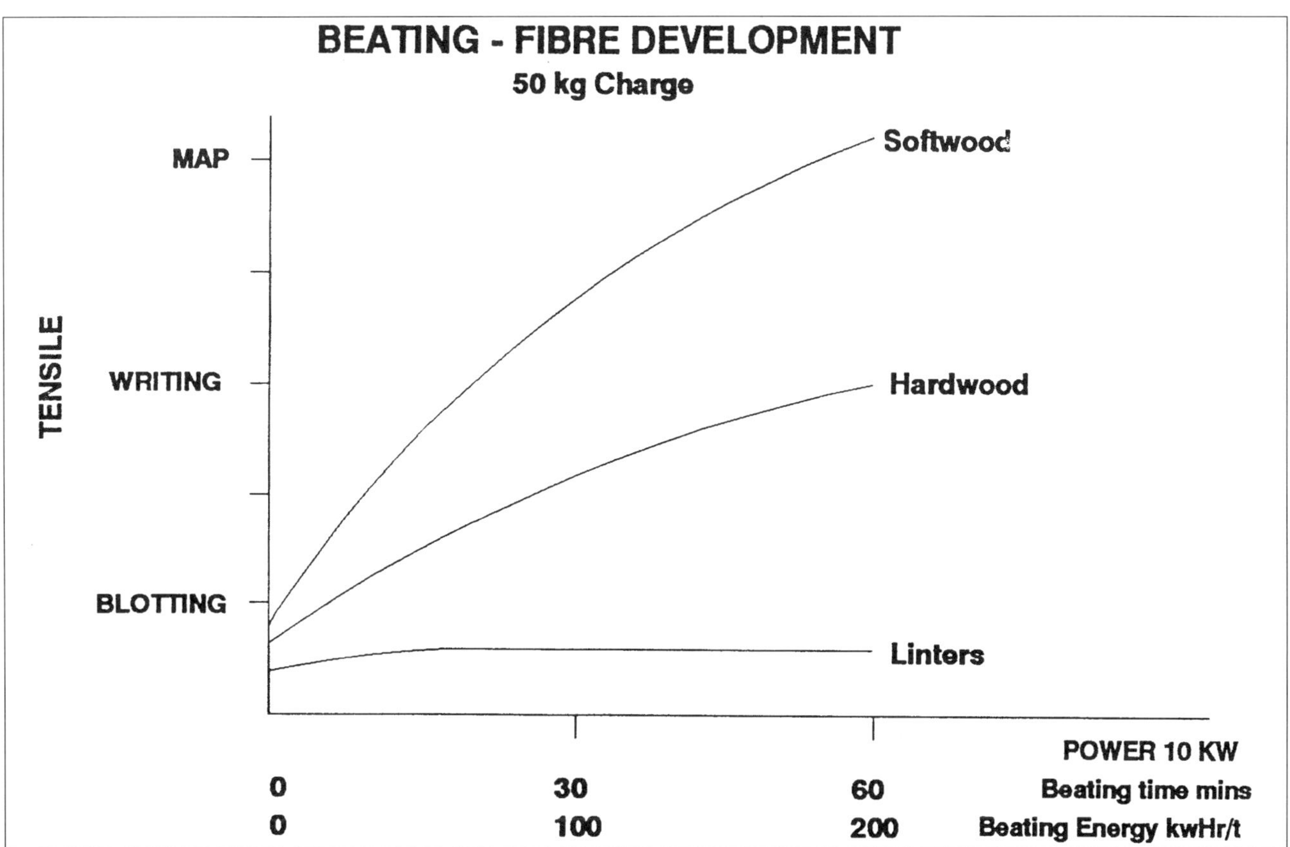

Figure 1: *Development of Tensile Strength with Beating Time*

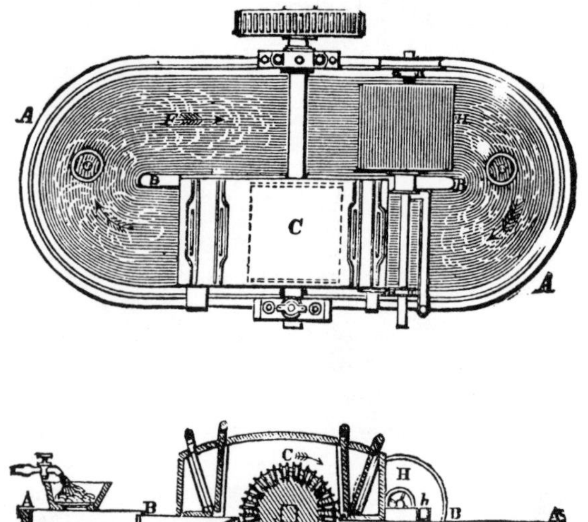

Figure 2: *Mid 19th Century Rag Engine*

than forty fold at 72°SR.[2] These fibrils also allow the fibres to bond together better giving a stronger sheet. The cutting and fibrillating actions take place at the same time in the beater, stamper or refiner. The skill comes in favouring one more than the other depending on the properties required, tear strength, tensile strength, bulk, porosity etc. See Figure 1. There are still mathematical theories being produced trying to rationalise the effects to make them predictable, but we generally find the Specific Edge Load theory the most practical.

Fibres are small compared with the equipment used to treat them.

	Length	Diameter
Cotton Linters	1 to 6 mm	20 - 30 μ
Rags	up to 30 mm	20 - 30 μ
Pine	2 - 3 mm	22 - 50 μ
Poplar	1.5 mm	25 μ
Straw	1.5 mm	13 μ
Eucalyptus	.05 - .1 mm	8 - 16 μ

Whereas beater bars are 6 to 15 mm or more! Their separation is 50 to 100 μ and they approach each other at 10 m/sec so the shear forces and turbulence are intense.

Stampers

Fibre treatment developed from rubbing the raw material between stones, using a pestle and mortar through to the stamping mill.

One of these units is still in use at the Richard-de-Bas Mill at Ambert in France. Two sets of three stampers hammer up and down in two basins with a volume of about 100 L, containing about 6 kg of fibre each. The action is one of bruising the fibres, or beating them. They need to be used continuously over long periods to produce fine papers - 12 hours is typical.

Kollergangs

At the beginning of the 17th century in the Zaan region of Holland, wind power was used. This was far less predictable than water and *kapperij* or chopping mills were more effective - there is still one at the "Schoolmaster" mill. The action is to cut the fibres which it does quickly compared with the stampers which shred as well as cut. This was followed by treatment in a Kollergang, or edge runner mill, which was originally used for grinding the wood used for dyes. This system however could only be used for coarse wrapping paper.

Kollergangs are very ancient pieces of equipment but continued in use up to the early 1980s in the Canson Mill at Annonay in France for making tracing paper. Their action is very much a rubbing one with little cutting - ideal for tracing.

Beaters

In the late 1600s the hollander beater came into being and many mills in the Zaan region were converted. In 1673 it was found that metal bars (a mixture of copper, brass and a little silver) were used for fine papers. Iron was not suitable.[3]

The Hollander Beater of 1779 is illustrated in a drawing in Diderot's *Encyclopédie* which shows a gear driven beater roll with a bed plate, that sits in a trough with a cover and washing screens.[4]

As it developed during the 19th century the Hollander beater was used for two distinct functions as a Rag Engine, Breaker or Potcher for washing and dispersing rags to produce half-stuff, and secondly as a Beater to develop the fibre.

The Rag Engine shown in Figure 2 is typical of the mid 19th century with a beater roll with its bars, working on a barred bedplate. The roll is gear driven and to keep the gears in mesh, that end of the shaft could not be moved much, so raising and lowering the roll from the roll end caused uneven wear. The 24" roll rotated at about 230 rpm pushing the rags and water up and over the *backfall* and round the *midfeather*.

The water sprayed over the backfall and hit screens, or *Chesses*, which allowed water and dirt through but kept the *stuff* in so it dropped back into the beater. This washing action could also be achieved with a drum covered in a wire mesh and fitted with baffles inside, which directed the water into a duct built in the midfeather.

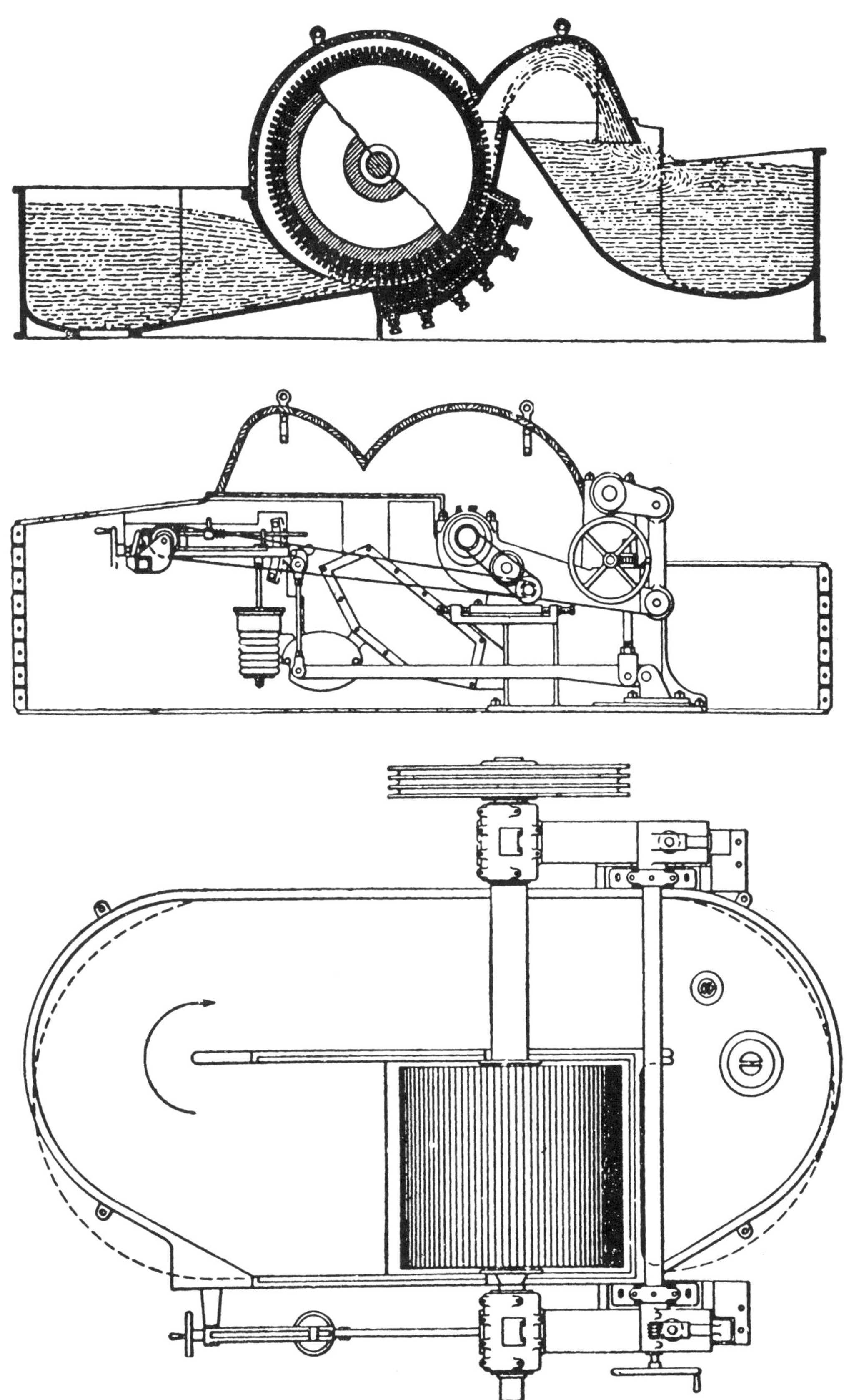

Figure 3: *Mid 20th Century Bertrams Beater*

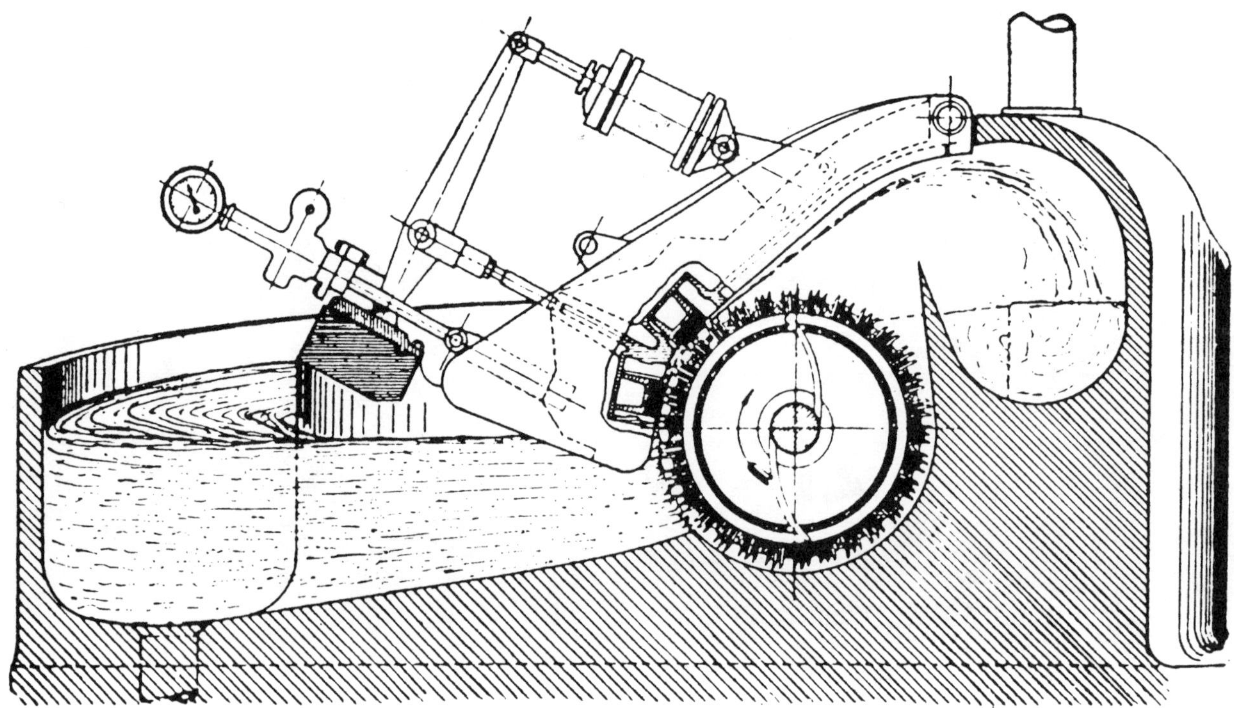

Figure 4: *Thorsen-Héry Beater — still in use at Chartham Paper Mill*

The bedplate consisted of a number of bars which from early days were laid at an angle of at least 3° to the beater roll.[5] This gives a scissor action; theoretical analysis and practical experiment early this century verified the improved action and much lower power consumption.[6] With the advent of electric motors to drive beaters, it became practical to monitor their actual power consumption.

G. & W. Bertram are credited with the introduction of the belt drive to beaters. This made it an easy matter to raise and lower the roll evenly. Bertrams became renowned for their beaters and developed them considerably over the years, in their booklet *A Study of the Beater* by Samuel Milne (3rd edition reprinted in 1939) they say their 360 lb *Hollander* beater produced 11,404,800 inch cuts/minute

(Bedplate 22 bars, roll 36" diameter, with 60 bars turning at 240 rpm), their latest 2,000 lb produced 230,000,000 inch cuts/minute and roll diameters were up to 72" and had automatic roll lowering. Bedplates covered 75° of the circumference of roll and could be pressed hydraulically against it, see Figure 3. For comparison a typical modern refiner runs at 162,800,000 inch cuts/minute.

Stoneywood Mill

Arjo Wiggins Mill at Stoneywood still has a number of Bertrams breakers and beaters in position though they have not been used since the late '70s. At that time their No.3 paper machine made 1 T/hr of paper using 9 washers and 9 beaters - usually though at least 1 pair was down for re-barring. All were made by Bertrams.

Figure 5 (right): *Different ways of circulating stuff around the beater.*
1. *Debié design (1868) — note small beater roll.*
2. *Strobach design — the Voith design was similar but the impeller was the other side of the mid-feather. Note the screwed impeller vanes.*
3. *Thomas Nugent's design in Whippany (1872) — note conical impeller.*
4. *Another Debié design (1868) using a screw and hydraulic bed plate.*
5. *The Acme design.*
6. *The Taylor design.*
7. *The Hemmer-Hollander 1894/95.*
8. *The Karger-Hollander 1893/94 — an Umpherston beater with Debié type of impeller*
9. *The Hromadnik-Ganzzenghollander (1895) — note the raised beater roll*
10. *The Kolliker-Hollander — here the propeller is after the beater roll.*

Crockett: The Development of the Beater

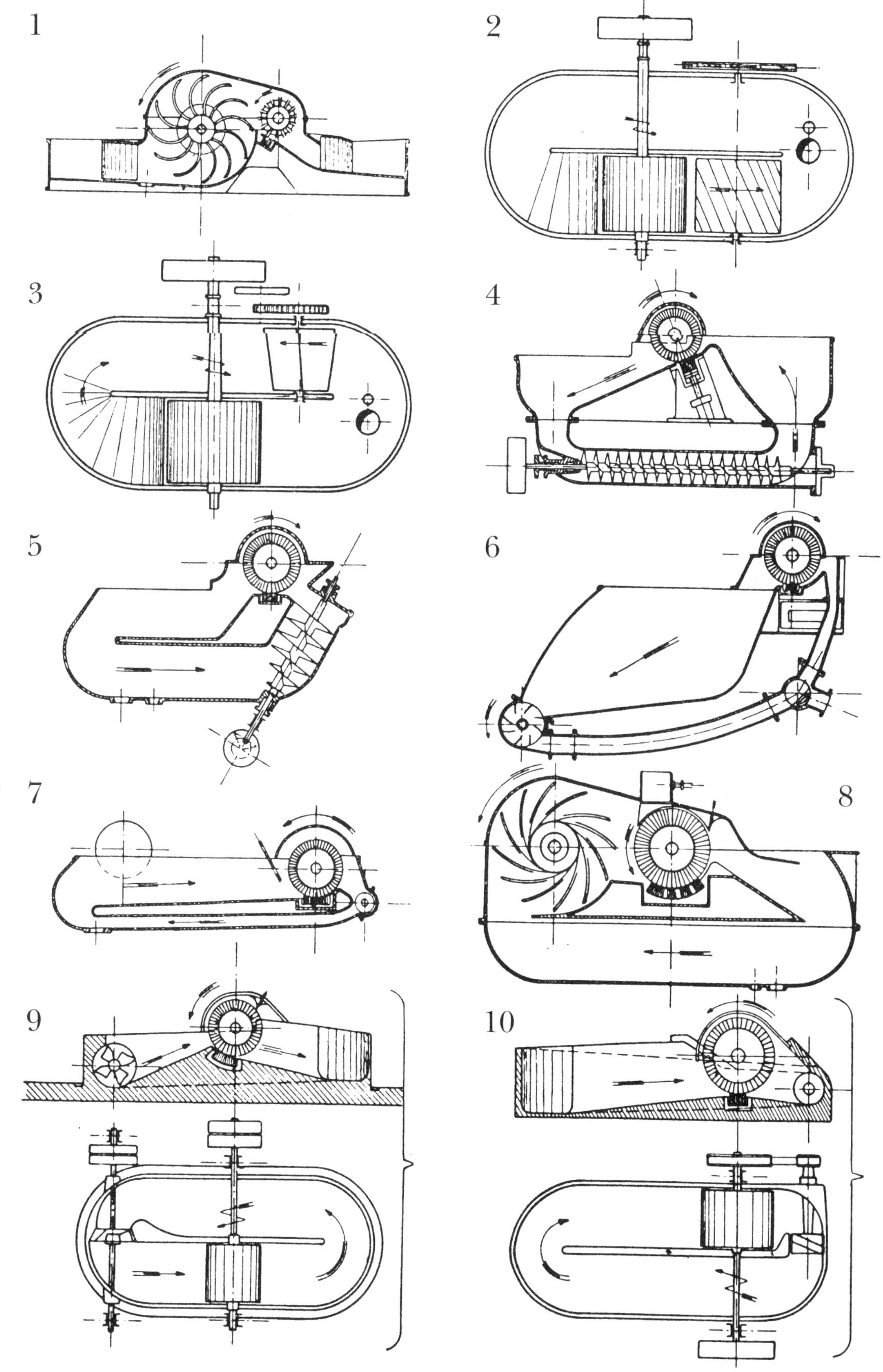

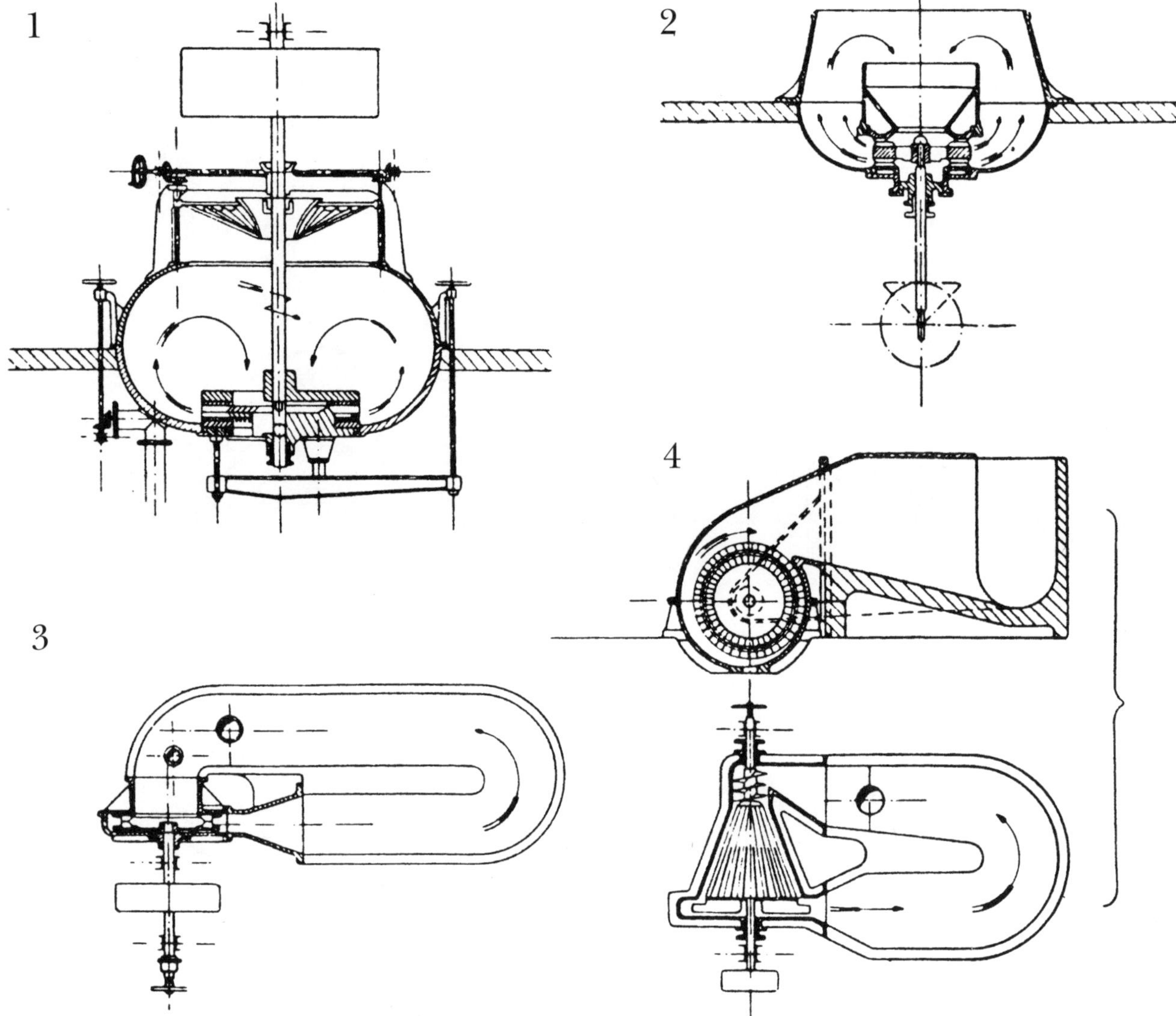

Figure 6: *Refiners that are* batch *run as beaters are: modern refiners usually run continuously.*
1. *The Gould Spherical beater, fairly widely used in 1870/72.*
2. *The Gould beater appeared in Germany as the* Shulte-Hollander *with the drive from underneath.*
3. *The Cook & Hibbert patent of 1887 with a Kingsland type of disc refining head.*
4. *The Axel Hisinger design with a conical refining head.*

The rags were taken in wheeled boxes by the *slide* men from the boilers and put into the Potchers - these were much larger than the beaters and only their foundations can be seen now. In the potchers the rags were washed and bleached, and then pumped into the steeps to drain over 2-3 hours. The steeps were concrete chests with perforated tiles in the bottom. The half stuff having drained in the steeps was then taken in boxes to the breakers (washers) which were filled with water. The beaterman slowly lowered the roll feeling the vibration in the bedplate with his foot. If the roll came down too fast the drive belt could be thrown, and since the engines were driven in banks of 4, all 4 had to stop till the belt was put back on. The breakers are 13' x 7'4" with 38" x 38" rolls - each contained 250 lb of stuff.

The half stuff was further broken down in the breaker and, if dirty or frothy from chemicals, was washed till the water ran clean, the water of course went straight to the river. After washing the half stuff went to the beater and was beaten to a specified drainage figure, °SR. The beaterman continually went round listening to the bedplates with a stick to check the power going into the stuff. Depending on the final product, the stuff could be diluted to give the beaters a greater cutting effect. The beaters are 10' x 5'7" with 33" x 33" rolls.

Typically 2 washers and beaters would be on linen, 3 on cotton and 2 on broke. Linen as it was beaten tended to feel slimy - the addition of rice starch had the same effect.

Crockett: The Development of the Beater

The manning on No.3 machine for 1 T/hr (or less) on a 3 shift operation was:

		Total
Day people in rag plant	5 women	7
	2 men	
Rag boilers	1/shift	3
Rag potchers	1/shift	3
Steeps	1/shift	3
Steeps to breakers and other duties	1/shift	3
Breakers	1/shift	3
Beaters	1/shift	3
		25

Now three machines make over 9 T/hour of paper and employ 2 men/shift on three shifts.[7]

Two Problems

The classic beater poses two problems, one is that the stuff near the midfeather goes round and past the beater roll more times than that on the outside of the trough. This gives uneven treatment and while many papermakers claimed a wide distribution of fibre length was good, others preferred more even treatment. This led to the Umpherston beater - made in the U.K. by Bertrams - which was patented in 1880. It passes the stuff back underneath the beater roll so all pulp is treated evenly.

There was a lively debate carried out in the letters of the *Paper Trade Review* in 1910 about the speed of beaters and the speed of circulation, this followed the publication in German of Clayton Beadle's book *The Theory & Practice of Beating*. It was also claimed that the Umpherston beater did double the amount of work for the same floor space as ordinary beating engines.[8]

The second problem with the beater is that the beater roll is used to pump the stuff round the beater as well as developing the fibres. As it was immersed in the stuff the turbulence and churning of the stuff by the roll wasted a lot of energy. Several designs got over this by separating the functions of pump and beater, such as the Bertrams Acme beater and Taylor tower beater by Masson Scott Ltd.

One beater which got over this problem elegantly and which is still in use to day at Chartham Paper Mills is the Thorsen-Héry beater, see Figure 4. Here the roll is rigidly fixed in position and the "bedplates" are lowered onto it; the roll turns in the opposite direction to normal and so is turning largely in air.

A sales letter of 1941 from Bentley & Jackson Ltd. says "As you will see (from the trial), the results obtained with our beaters are greatly superior to those given by the Bertram beater and this latter, as you know, is considered as being one of the best in the world."

Chartham have 4 size 2 beaters on their No.1 machine and 2 on No.2. They are fitted with basalt lava blocks and bedplates for making tracing paper and take 600 lb of pulp at 6%. They are run to a set power consumption, adjusted manually, for a standard time, depending on product, of between 30 minutes and 3 hours. They were installed just after the War.

Variations in Design

Die Papierfabrikation und duren Maschinen by Friedrich Müller gives some interesting variations on the design of beaters. Figures 5 and 6 show some which are getting very close to being totally enclosed refiners, the equipment that has since the War brought the 300 year reign of the Hollander to an end.

In 1910 in the *Paper Trade Journal* they reported that:

> A discussion has been carried on as to who is the most important person in papermaking. Some of the writers are inclined to give credit to the machineman, but many favour the beaterman. In reference to the latter one correspondent says: "The beaterman is the most important, for, as we all agree, it is the beating room where the paper is made, as no matter what kind of a machine tender you have he can't make unless he gets his stuff right for that particular grade he is going to make, as he can't shorten or lengthen out his fibres. When once the beaterman puts it down into his chest that is the end of it, as it has got to be right, as there is no getting it back again, or half of his stuff goes down the sewer. As for the steam engineer, he has mechanical appliances for his work, and if he has enough his work comes easy, but the beaterman has got to use judgement, keep all his strength in the fibres, look after the colour, and many a time to keep three machines going with stuff all making different grades of paper, and has to be on the look-out all the time if he wants his work to come out all right.

References

1   Clapperton, R.H. *Modern Papermaking 3rd Edition*, p.140.
2   Emerton, H.W. *Composition and Structure of Papermaking Fibres and the Effect on these of the Beating Process.*
3   Cottrall, L.G. *History of the Development of Beating and Beating Plant.*
4   Hills, R.L. *Paper Making in Britain 1488-1988.*
5   Balston, J.N. *The Elder James Whatman Vol.2, p.231.*
6   Smith, Sigurd *The Action of the Beater, p.173.*
7   *Conversation and correspondence with Lewie Stewart, Production Manager, Stoneywood Mill.*
8   Watt, Alexander *The Art of Papermaking, p.106.*

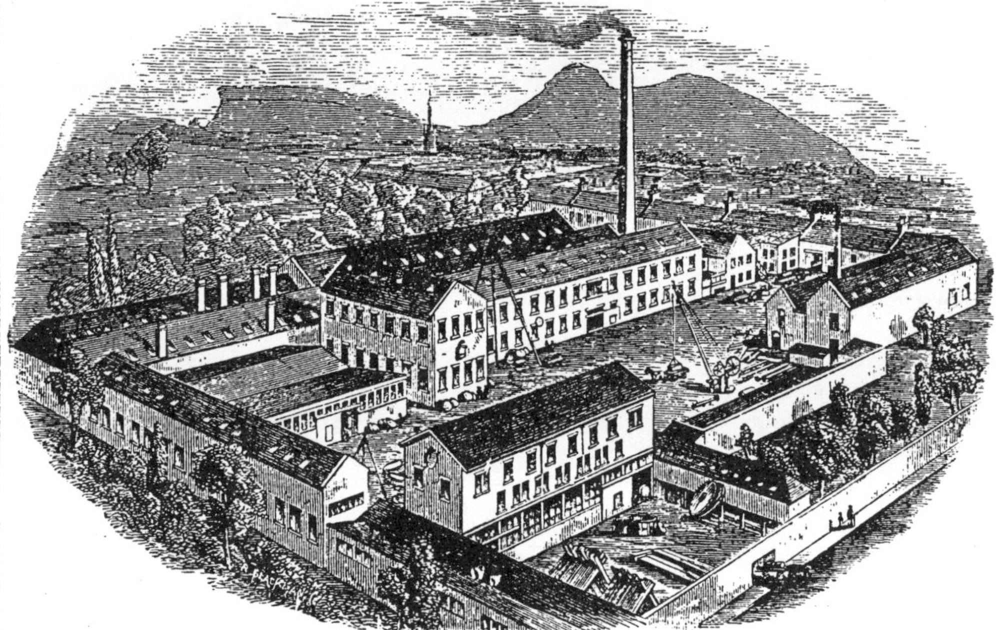

ESTABLISHED 1828.

Geo. & Wm. BERTRAM,

ENGINEERS,

ST. KATHERINE'S WORKS, SCIENNES,
EDINBURGH.

MANUFACTURERS OF ALL KINDS
OF
MACHINERY
FOR
PAPER-MAKERS.

PLEASE NOTE THE ADDRESS.

THE BALL FAMILY, PAPERMAKERS OF SURREY AND NORTHERN FRANCE

Alan Crocker and Anne Phillips

1. Charles Ball I and Charles Ball II in Northampton
Rush Mill; 1763-1790

During the late seventeenth century many skilled paper-makers, particularly Huguenots, left France to settle in England, where their expertise provided the foundations for great improvements in the quality of paper being produced in English mills.[1] The present article gives an account of a successful English papermaker Charles Ball and his son Charles Ashby who, in the early nineteenth century during the economic depression which followed the Napoleonic Wars, moved in the opposite direction. They were able to establish several mills in France and to perfect various improvements in the manufacture of paper. Ball's father and grandfather, again named Charles, were also papermakers and the locations at which the family were active in the two countries are indicated in figure 1. A selective family tree is given in figure 2.

In 1767 Charles Ball I, who since 1763 had been the tenant papermaker at Rush Mill on the River Nene in Northampton, died.[2,3] In his will[4] he left most of his property to his son Charles, who appears to have been baptised in 1761 at Farningham in Kent.[5] This is near Eynsford Mill on the River Dart where his father may have been working. However there are no records of papermakers with the surname Ball being in mills in Kent at this time.[6] At his death Charles Ball was presumably still a young man as his children were all minors, his wife was pregnant and his father, also called Charles, was still alive.[4] His widow became the tenant of the mill but by December 1769 this had been destroyed by fire and a papermaker called Francis Hayes took a lease on the land and rebuilt the mill. He made good quality white paper and his family continued at the mill until 1811 when John Hayes was bankrupt.[2] Later Rush Mill became famous for producing in 1840 the paper with the security "crown" watermark on which the Penny Black, the first adhesive postage stamp in the world, was printed.[7]

Meanwhile in 1780 Charles Ball II, claiming to be 21 years old, was married in St. Clement Danes church in London to Martha Weaver, of the parish of St. Martin in the Fields, who was 19.[8] Their first two children Charles and Edmund Richard, were baptised in Northampton in 1783 and 1786[9] so Ball was probably working at Rush Mill with Francis Hayes. The name Edmund was uncommon at this time so it is likely that Ball's second child was named after a relative. This could have been Edmund Ball, who was the papermaker from 1761 to 1778 at Ball's or Frog Mill at West Wycombe in Buckinghamshire.[10]

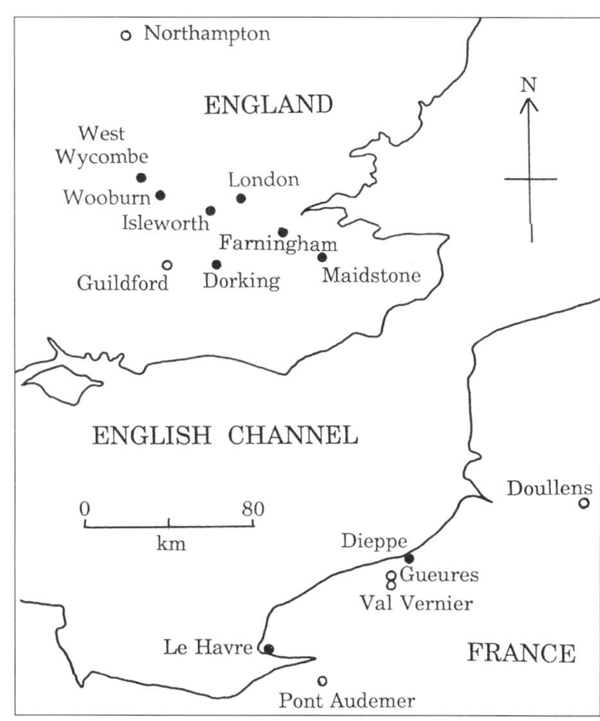

Figure 1: *Location map of towns in England and France referred to in the text. The sites of paper mills definitely operated by the Ball family are represented by open symbols. The five sites near Guildford are shown in detail in figure 3.*

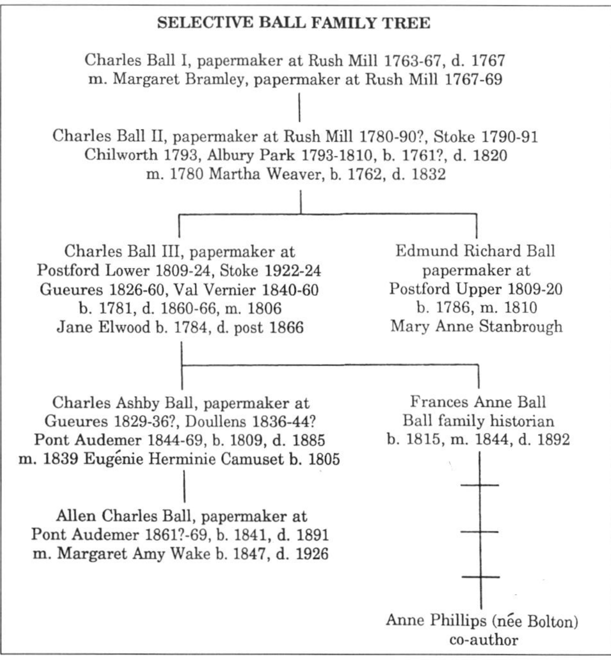

Figure 2: *Selective Ball family tree.*

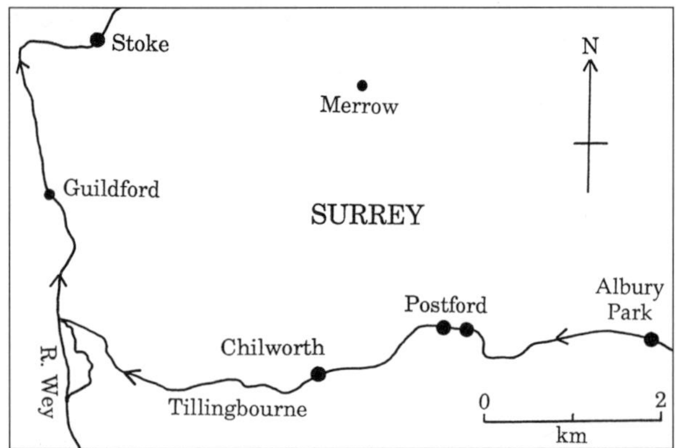

Figure 3: *Location map of Stoke, Chilworth, Albury Park and Postford Mills near Guildford in Surrey which were operated by the Ball family.*

The next we hear of Charles Ball II is in 1790, when he reappears at the paper mill at Stoke near Guildford, in Surrey.

2. Charles Ball II in Surrey

Stoke, Chilworth and Albury Park Mills; 1790-1810

Between 1790 and 1810, when he retired, Charles Ball II was the master papermaker at three mills near Guildford in Surrey. These were Stoke, Chilworth and Albury Park mills,[11,12,13] the locations of which are indicated on the sketch map of figure 3. Stoke Mill on the River Wey is the Surrey paper mill for which the earliest documentary information is available.[11-13] It was established shortly before 1635 and paper was made at the site until 1869. Charles Beauties sketched the mill in about 1800.[14] In 1786 John Grove, the papermaker at Glory Mill at Wooburn in Buckinghamshire (see figure 1) took out a 21 year lease on the mill. However the land tax returns show the following occupants: 1786-9, Joseph Callow; 1790-1, Charles Ball; 1792-3, Elsee and Cotton; 1795-1823, John Grove.[11,13] Callow was the papermaker at Chilworth Mill but by 1791 he was bankrupt.[12] Elsee and Cotton were probably stationers in London and Bristol respectively.[15] It is also interesting that the Stoke land tax returns for 1790 and 1793 were recorded on paper bearing Charles Ball watermarks, an example being given in figure 4.[16] In addition Ball's son Frederick was baptised at Stoke on 12 February 1792.[17] It is clear therefore that for a few years after he left Rush Mill at Northampton Ball was active at Stoke.

In 1793 the partnership between Charles Ball and William Wilcox, papermakers at Chilworth, was dissolved. A paper mill had been established at Chilworth on the Tillingbourne, a tributary of the River Wey, in 1704. By 1728 there were two mills at the site, known as the Great Mill and the Little Mill, and both continued until about 1830. The Little Mill then closed but the Great Mill survived until 1870. Captain Wilcocks [Wilcox] paid the land tax for the sites in 1792-3 but nothing more is known of him.[12] However it is interesting that Ball's

younger daughter Caroline named her son Francis Wilcox.[18] Unfortunately there are no further references to Ball working at Chilworth; his association with the mill was again short lived.

From 1794/5 to 1810 Charles Ball II paid the land tax for Albury Park Mill, which was on the Tillingbourne upstream from Chilworth (see figure 3).[12] Previously this had been a corn mill and it appears that Ball built the paper mill. This is shown in a sketch of 1800[19] and an engraving of the mill reproduced elsewhere[12] which bears the caption "Albury Mill & Paper Mould Manufactory near Guildford in Surrey". This was engraved by Harry Ashby, a London printer and engraver specialising in bill-heads, banknotes and specimen sheets of calligraphy, who was a business partner of Ball.[20,21]

In 1793 or 1794 a stranger called at the mill and asked Charles Ball to make some paper containing particular watermarks.[22] The order was repeated frequently but sometimes the watermarks had to be modified. Later it was discovered that the mysterious visitor was the Count of Artois, afterwards Charles X, King of France and Navarre, and that the paper was for false assignats. The changes in the watermarks were necessary because when the officers of the Republic discovered the forgeries, they altered the form of the assignats. Hence Charles Ball played a role in the attempt of the French royal family to undermine the Revolution by forging assignats.

It was a period when many master papermakers were having problems with their workers and Ball was active in meetings of the employers. For example with other papermakers at mills near Guildford he signed the following letter to Nathaniel Davies, the solicitor employed by the Committee of Master Paper-Makers, which held their meetings at the George and Vulture Tavern in Cornhill, London.[23]

> *Mr Davis, Sir*
> *Guildford, March 26, 1796*
> *In consequence of what passed at the meeting of the Master Papermakers at the George and Vulture on Tuesday last, We of that Branch in the Neighbourhood have had a Meeting when it was unanimously agreed to give our Men a fortnights warning from this day, unless they comply with all the terms required and agreed to at the Second Meeting namely "to desist from giving any assistance to those Men who have continued against their Masters for an advance of Wages" and also to "desist from what is called their Turns to the Journeymen at present out of work" till they return to their Duty. We hope these measures may prove efficacious and beg of you to lay these our Resolutions before the Committee.*
> *We remain Sir, Your humble Servants*
> *Edward Hughs, Chas Ball, Smith & Knight, John Groves, Thomas Harrison*

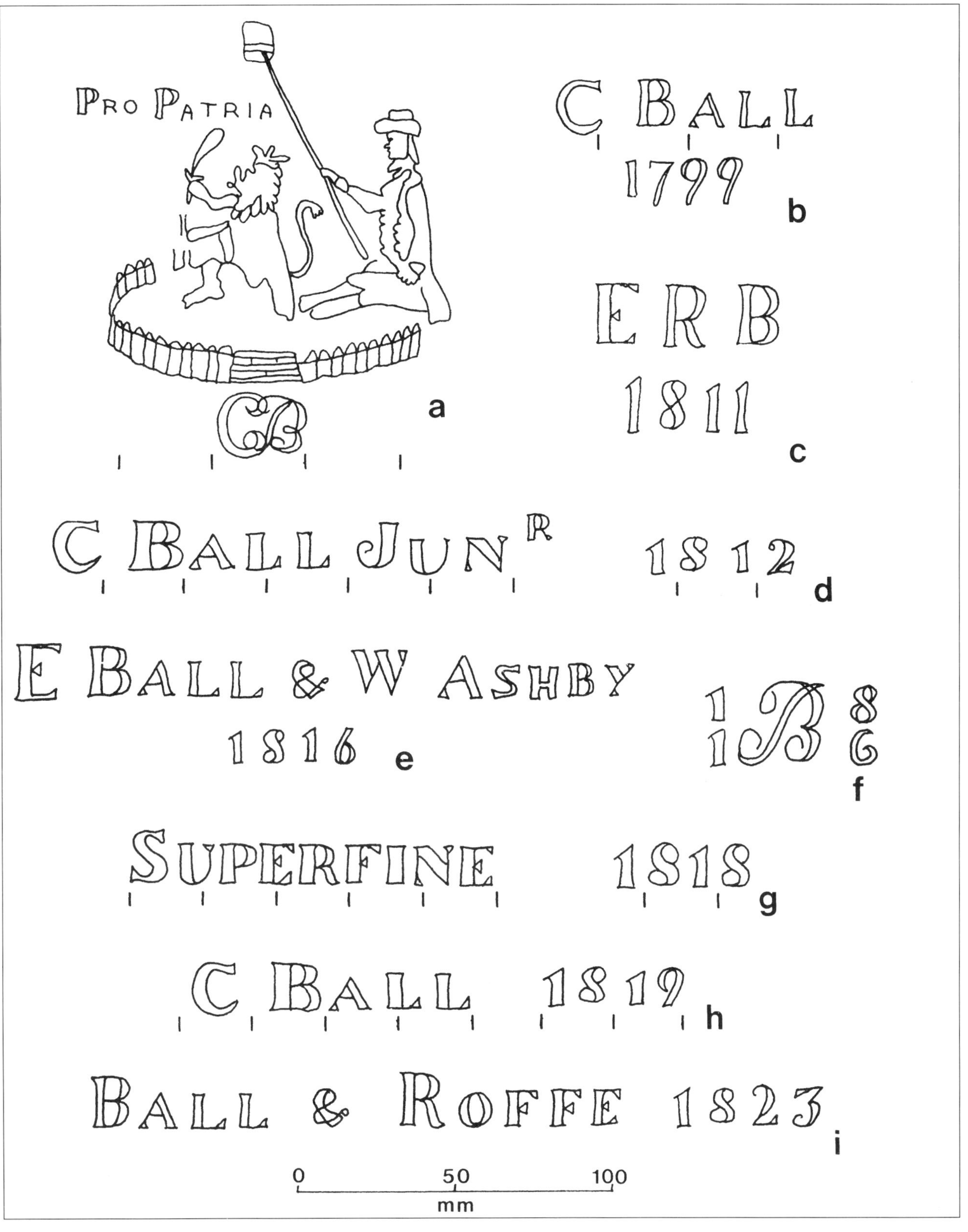

Figure 4: *Watermarks in paper made by the Ball family in Surrey. (a,b) Charles Ball II, (d,f,h,i) Charles III, (c,e,g) Edmund Richard Ball (e,g) William May Ashby and (i) Charles Roffe at (a) Stoke Milll (1792), (b) Albury Park Mill, (c,e,g) Postford Upper Mill, (d,h) Postford Lower Mill and (i) Postford Lower or Stoke Mill. The chain line spacings of the laid paper are as indicated, the spacing of the laid lines being about 1.05 mm in each case.*

Figure 5: *Portrait of Jane Ball (née Elwood), wife of Charles Ball III.*[35]

Within two months of this letter being written a Bill which aimed to prevent unlawful combinations of workmen employed in paper manufacture was passed by Parliament.

Ball's daughter Caroline Martha was baptised at Albury in 1800[24] but from 1803 to 1816 he lived in the adjacent parish of Merrow.[25] His son Edmund Richard was married there in 1810 and Caroline Martha in 1815.[26] By 1818 Ball was living at Clapham, now in south west London,[27] and he died aged 58 in Lombard Street in 1820. He is buried in a large tomb in Merrow churchyard together with Caroline Martha and her husband Francis Collins, both of whom died in 1828, their daughter Ellen and their son Francis Wilcox, who is referred to above.[18] The tombstone also records that Ball's wife Martha died in London in 1832. His obituary stated that he was "eminent as the inventor and manufacturer of superior bankers' note paper, and late of the firm of Ball & Ashby, engravers etc".[21]

Prior to the arrival of Charles Ball II at Albury Park Mill the manor had been held by Admiral William Clement Finch who harassed the villagers living near the church and manor house so much that many of them moved to the hamlet of Weston Street 1.5 km to the west. Finch died in 1794 and during Ball's occupation of the paper mill the village remained stable. In particular from 1800 to 1810 the manor was owned by Samuel Thornton,[12] an East India merchant and a governor of the Bank of England, with whom Ball was friendly.[28] There is no evidence however that Ball produced paper for the Bank. The next owner was Charles Wall whose arrival coincided with Ball's departure in 1810. He demolished most of the remaining cottages and the mill closed. Eventually in about 1850 it was replaced by a mock Tudor laundry for the nearby mansion and this has been converted into an attractive house.[12]

3. Charles Ball III and Edmund Richard Ball in Surrey
Postford Lower, Postford Upper and Stoke Mills; 1809-1824

In 1809 Charles Ball II took out leases of 61 years on two newly constructed paper mills at Postford which, as shown in figure 3, lies on the Tillingbourne between Chilworth and Albury.[12] He installed his eldest son Charles at Postford Lower Mill on Postford Pond and his younger son Edmund Richard at Postford Upper Mill on Paynes Pond, later to be renamed Waterloo Pond. Charles Ball junior settled in the newly built Postford Hill, now known as Postford House, and Edmund Richard in a cottage residence adjacent to the Upper Mill. When the Chilworth Manor estates were sold in 1813, each of the mills had three vats and overshot water-wheels 11 feet in diameter. The combined annual rent was £168. Then in 1815 Edmund Richard Ball was joined by William May Ashby, presumably a relative of Harry Ashby the engraver referred to above. Also at this time Charles Ball III dropped the "junior" from his title. These changes are reflected by watermarks which are illustrated in figure 4.[12,16,29] An engraving, published in 1832, shows the Lower Mill at about this time.[30]

There are many records which demonstrate that the Upper and Lower mills, which were allocated excise numbers 386 and 387 in 1816, made banknote paper.[12] For example in about 1818 Charles Ball III submitted to the Bank of England a sample of his paper "manufactured solely from a vegetable substance which can be rendered opaque and watermarks of any description can be introduced into it".[31] Also an 1819 letter from Ball & Ashby to William Henry Roberts, a representative in Scotland, concerns paper for the Bank of Ireland, the British Linen Bank, the Royal Bank, the Aberdeen Bank and the Perth Bank.[29] It includes the following invoice:

Perth Bank Company.
To Ball & Ashby

Oct 23rd		£	s	d
	10 small paper 24s	12.	0.	0
	9 large ditto 24s	10.	16.	0
	2 new moulds	10.	10.	0
		33.	6.	0

It was at about this time that the infant Martin Tupper, who was to become a very popular Victorian versifier, whose name is now synonymous with pretentious twaddle, was taken to the mill by his nurse. While she dallied with one of the papermakers he amused himself by making banknote paper.[12] Then in 1822 the Radical political and financial campaigner William Cobbett, while on a visit to Chilworth and Albury, recorded his celebrated diatribe against "two of the most damnable inventions that ever sprang from the minds of man under the influence of the devil! namely, the making of "gunpowder and of banknotes!" He hated paper money and deplored the fact "these springs should be perverted into means of spreading misery over a whole nation".[12]

Following the Napoleonic Wars many banks failed so that banknote papermakers were in difficulties. In particular in 1820 Edmund Richard Ball and William May Ashby became bankrupt and the Upper Mill, the associated residence and their contents were auctioned.[12,32,33] The mill contained about 1000 reams of banknote paper suitable for the Irish, Dundee, Isle of Wight and other banks, 115 reams for Irish linen and 100 reams of foolscap. The equipment included 4 iron screw presses, large coppers, cast iron stoves, work benches, stages, drying rooms and frames. It was taken over by Hugh Rowland the papermaker at Chilworth Mill but he left in 1824 to be replaced by John and Charles Francis Hayes.[12] This provides an interesting link with Rush Mill, which the Hayes family had taken over following the death in 1767 of Charles Ball I. However in 1826 they also became bankrupt. Meanwhile at the Lower Mill, now with four vats, Charles Ball III became bankrupt in 1821 and the mill and Postford Hill were for sale. However Ball continued at the mill with a new partner Charles Roffe and a year later they were also at Stoke Mill, where Charles Ball II had been in the early 1790s. Some Ball & Roffe watermarks are reproduced in figure 4. However in 1824 Roffe became the sole occupier of the two mills and a year later he too was bankrupt.[12]

Charles Ball III had married Jane Elwood of Capel, 9 km S of Dorking in Surrey (see figure 1), 1806[34] and her portrait is shown in figure 5.[35] They had nine children between 1807 and 1827, including Charles Ashby born in 1809 and Frances Anne, the great-great-grandmother of one of the authors of this paper (AP), born in 1815.[36] Edmund Richard Ball had married Mary Anne Stanbrough of Isleworth in Middlesex in 1810.[26] Their eldest child was baptised in Guildford in 1812 and a further five children in London in 1822.[37] This was at St. Edmund King and Martyr in Lombard Street, where the Balls were living and where Charles II had died in 1820. The story continues however with Charles Ball III who following his departure from Postford Lower and Stoke mills seems to have moved briefly to Dorking (see figure 1), where his youngest child was born,[36] before emigrating to become a papermaker in France.

4. Charles Ball III and Charles Ashby Ball in Normandy and Picardy
Gueures, Val Vernier and Doullens; 1826-1860
In December 1826 Charles Ball III was a partner in the firm Gosse, Ball & Muller, tenants of a new paper mill at

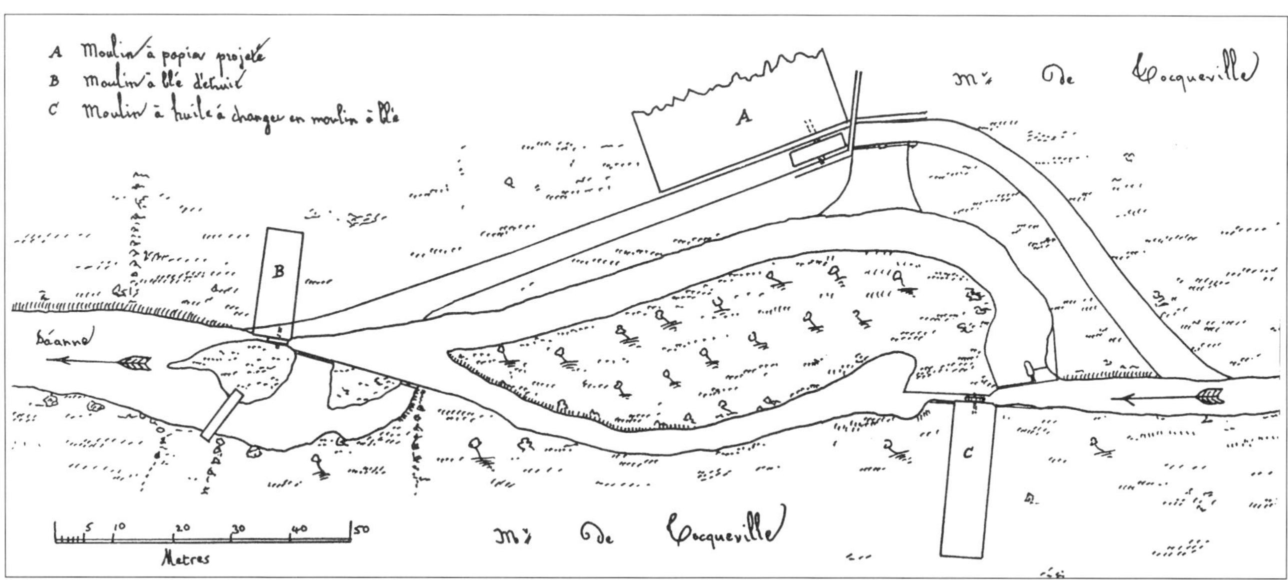

Figure 6: *Plan of part of the property of the Comte de Tocqueville on the River Sâane at Gueures with north approximately to the left. It shows (A) a proposed paper mill, (B) a demolished corn mill and (C) an oil mill. A new water channel is indicated leading to the paper mill. Based on an original dated 9 September 1826.*[39] (Drawn by Alan Crocker).

Gueures, 13 km SW of Dieppe in Seine Maritime, Normandy (see figure 1).[39] This mill was established by a wealthy and enterprising anglophile, Pierre Victor, Comte de Tocqueville, who was married to London-born Anne Tulloch.[40,41] It became the focus of his endeavours to introduce industry based on English practice into a rural village community of fewer than 600 people. The mill was located on the right bank of the River Sâane, less than 100 m from his chateau and, as shown on the 1826 plan of figure 6, involved the closure of a corn mill, the conversion of an oil mill into a corn mill and the creation of a new mill stream.[39] It started production in 1827 and had one of the four Donkin papermaking machines installed in France at this time.[41,42] In 1829 it had 120 employees, many of whom came from England, and produced 100 tonnes of paper annually, most of which was sold in Paris and Rouen.[41,43]

Charles Ashby Ball is said to have entered the paper business in 1826 and purchased the Gueures works with his father in 1829.[44] In addition, in an account of her family written in about 1887, his sister Frances Anne stated that when her father left Postford and Stoke he went to France and under the patronage of the Comte de Tocqueville built a paper mill at Gueures, where

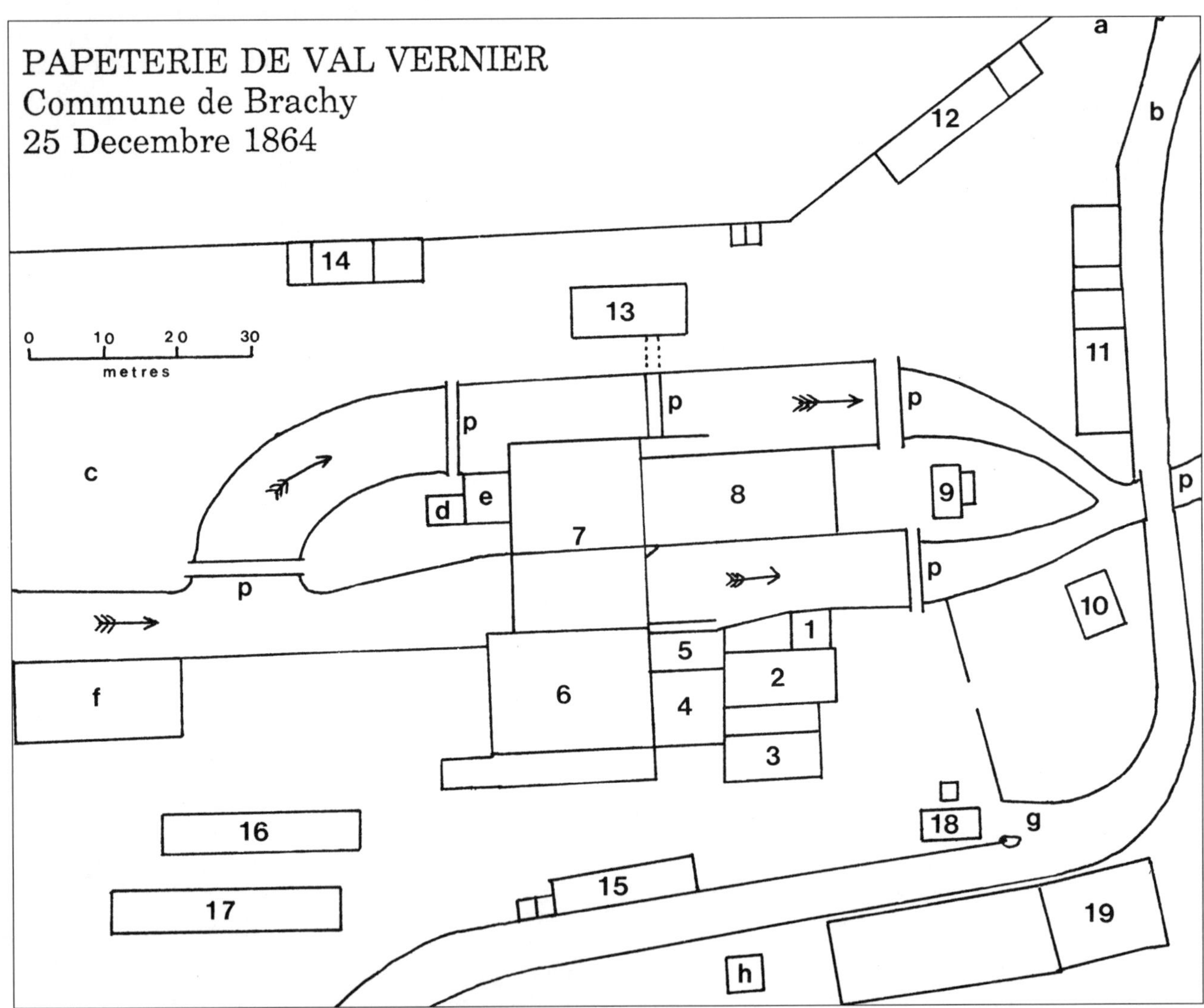

Figure 7: *Plan of the paper mill on the River Sâane at Val Vernier, based on an original dated 25 December 1864.[46] In translation the labelling is as follows: (1) gas, (2) heaters, (3) steam engines, (4,5) —, (6) bleaching and washing, (7) waterwheels and refiners, (8) papermaking machine, (9) office, (10) director's house, (11) lodgings, caretaker and store, (12) paper store, (13) store, (14) stable and store, (15) store, (16-17) sheds, (18) weighbridge, (19) first and second evaporation units, (a) main entrance, (b) road to Gueures, (c) straw store, (d) forge, (e) tower, (f) chlorine preparation, (g) entrance, (h) chimney, (p) six bridges. A few of these buildings, including (9) and (10), survive but most were replaced in the late 19th century or more recently.*

Charles Ashby began his career.[28] The fact that Charles Ball was not present at the wedding of his eldest daughter at Dorking in Surrey in March 1828 also suggests that he was busy in France.[38] The mill flourished and de Tocqueville built terraces of industrial housing for the workers and a pair of picturesque "English chalets" at the entrance to his park.[41] By 1838 the workforce had increased to about 200 and a new initiative to produce machine-made wallpaper was launched. Capital was raised and a new machine installed but the venture was unsuccessful. It ceased in about 1841 and following the death of his wife de Tocqueville left the area.[41,43,45] Throughout this period the mill was operated by Muller & Co. in association with Bouchard-Oudin, Isadore Drouard, a wallpaper manufacturer of Paris, and Gosse de Serlay, an architect of Rouen.[41,43] However it appears that the Ball family were still involved as in 1858-60, Ball & Co. were stated to be the owners. Ferdinand Matthias & Co. then took over and closed the Gueures works in 1861.[46,47]

The main paper mill building at Gueures disappeared many years ago and even the former rag house, which survived until the 1980s, has now gone. However the manager's house, known as *La Source*, a single storey building thought to have been used as an office, much of the *cité ouvrière*, one of the *chalets á l'anglaise* and the chateau still survive.[45]

Meanwhile by 1837 Muller & Co. had acquired an oil mill and a corn mill at Val Vernier, near the village of Brachy, 2 km farther up the Sâane from Gueures.[41,46] In 1840 they were authorised to convert these mills into a paper mill, subject to several conditions on satisfactory control of the river water. They built a factory which made paper entirely from straw and in the 19th century was the only one to do so in this part of France. Technically it was very successful and in about 1845 the water power was supplemented by steam. However problems soon arose with the neighbours, those upstream complaining about irrigation of meadows and those downstream about the discharge of more or less acid water used in processing the straw. After 1845 this conflict became persistent but papermaking continued. Thus in 1855 Ball the elder of Val Vernier was selected to display his paper at the *Exposition Universelle* in Paris.[48] This must have been Charles Ball III who was then 74 years old. Also in 1858 and 1861 the mill was operated by Ball & Co.[45,47] Indeed it seems likely that the Ball family had been involved from the beginning of the venture as Charles Ashby Ball was said to have purchased the Valvennes (sic) works near Dieppe and manufactured paper from esparto and bleached straw.[44] However in 1860 Ferdinand Matthias & Co. had an old paper mill at Gueures and a more recent one at Val Vernier.[46] These changes in the title of the company did not however

solve the pollution problem and, following an enormous conflict in 1863/4, Matthias became bankrupt and Val Vernier finally closed in 1868 throwing some 80 people out of work.[46,47] A plan of the factory in 1864 is shown in figure 7.[41,46]

The mill was later used for weaving and bleaching thread. This industry gradually expanded and was well known for producing good quality cloth with the label *Fleur Bleue*. Finally it was closed because of competition from synthetic cloth. The site, retaining many of its late 19th century buildings but only a few from the paper-making period, is now occupied by a manufacturer of electrical materials.[47]

In 1836 Charles Ashby Ball, whose portrait is shown in figure 8[49], is said to have bought the paper mill at Doullens, 30 km N of Amiens in Somme, Picardy (see figure 1).[44] Three years later he married Eugénie Herminie Camuset, a cousin of the famous French chemist Bertholet. They met at Gueures, she became a protestant and the wedding took place in the Anglican church at East Malling, near Maidstone in Kent, a major papermaking centre.[28] It is interesting that one of the witnesses was Charles Roffe, formerly a partner of Charles Ball III. However the space of the marriage certificate for the name of the father of the bridegroom is left blank.[50] This might suggest that Ball had died but this is inconsistent with him exhibiting in Paris in 1855.[48] He is said to have "died near Havre where he went on account of his health"[28] but no record of his death at Le Havre has been found.

Charles and Eugénie Ball had a son, Allen Charles, who was born at Doullens in 1841. The birth was registered on 30 August, the father being described as a paper-maker living at St. Sulpice les Doullens.[51] The paper mill was on the site of a priory which was suppressed at the end of the 17th century although the chapel survived until the Revolution. The property was then sold and a corn mill erected on the right bank of the River Authie 2 km upstream from the town.[52,53,54] This was acquired by Auguste Darras who in 1822 requested permission which was granted a year later, to convert it into a paper mill.[55] This seems too early for it to have been a machine mill and indeed in 1838, after Charles Ashby Ball had arrived, it was necessary to erect new buildings, presumably for a papermaking machine, and the foundations revealed a fine mosaic pavement from the former priory.[52,53]

It is not known how long Ball stayed at Doullens but it seems likely that he sold his interest in the mill in 1844 when he purchased another mill at Pont Audemer.[40] He does not seem to have been associated with a second paper mill at Marais Sec 1 km N of Doullens. This was on the Grouches, a tributary of the Authie, and was estab-lished by Lemoine sometime in the 19th century.[52] The

St. Sulpice mill flourished under the direction of Horne in 1838, Raux-Grimand in 1865, Maisonneuve in the 1890s and then the Soustre family. It was operated by *Cenpa* after the war, cardboard box manufacture being introduced in 1950, and then in 1960 *Cenpa* merged with *La Rochette*. The St. Sulpice works closed in 1976 when *La Rochette-Cenpa* opened a new box factory on the Doullens industrial estate. However the old buildings survive and are used to manufacture electrical goods.[52-54]

5. Charles Ashby Ball and Allen Charles Ball in Normandy

Pont Audemer Mill; 1844-1869

In 1838 *La Compagnie des Etablissements de la Risle* was formed and developed an industrial complex at Pont Audemer, which is in Eure, Normandy, 32 km ESE of Le Havre (see figure 1). The factory, called *La Madeleine*, was on the right bank of the Risle, immediately downstream from the town and 12 km above its confluence with the Seine. It consisted of a paper mill, an iron foundry, a zinc plating works, a corn mill and a tanyard. The power was provided by constructing a timber weir across the river, which diverted water into a leat, which fed waterwheels in the works.[56] An attraction of the site was that the Risle was navigable below Pont Audemer and there were no towns downstream to complain about pollution. The paper mill, which had two machines, was purchased by Charles Ashby Ball and opened in 1844.[44,56] It was very successful and whilst there Ball developed methods of producing paper from esparto grass and straw.[44] For example in 1855 M. Ball & Cie of Pont Audemer were selected to exhibit at the *Exposition Universelle* in Paris and were awarded a medal for a display of diverse papers remarkable for their quality and presentation. The exhibition catalogue indicates that the exhibitor was Charles Ashby but it is thought that his 74 year old father was also involved.[48,56]

The only known illustration of the works is reproduced as figure 9.[57] This was probably drawn by a member of the family and appears to be a view from a window of their house looking across a terrace and garden towards a range of buildings, assumed to be part of the paper mill. Some of the garden trees appear to be willows suggesting that the location is indeed the moist valley bottom adjacent to the factory. The skyline indicates that the view is approximately towards the NE. However no remains of these buildings have been found.

Charles Ashby's sister Frances Anne recorded that he was always travelling about and went to Scotland to visit the paper manufactories. He became friendly with the Collins family, papermakers of Kelvindale Mill at Glasgow[28] which concentrated on esparto papers.[58] This led in 1869 to the marriage of his younger son Henry William, who was born in Pont Audemer in 1847, to Mary Anne, the daughter of Joshua Heywood Collins and

Figure 8: *Portrait of Charles Ashby Ball.*[49]

Figure 9: *Drawing labelled 'Pont Audemer' which is assumed to show part of the paper mill as seen from the Ball residence.[58] The skyline indicates that the view is from the south west.*

brother of Edward Collins junior. Frances Anne also noted that another nephew, Willie Crichton, married Georgina Hollingworth, a member of the well-known papermaking family at Maidstone in Kent. She also stated that Ball's wife Eugénie was very charitable and did much good among the workers.[28] Many of these probably lived in the surviving terraces of industrial housing lining *Rue de la Madeleine*. In 1866 Charles Ashby's elder son Allen Charles, born at Doullens in 1841, was said to be unmarried and associated with the direction of the paper mill.[56] Later he married Margaret Amy Wake, daughter of Sir William Wake, 11th Baronet of Courteenhall, Northampton, providing an interesting link with the county where his great-great-grandfather was a paper-maker.

In 1867 the Normandy Association for Progress of Agriculture, Industry and Arts, of which Charles Ashby was a member, held its annual meeting at Pont Audemer and visited the paper mill.[56] They described it as one of the most complete and interesting establishments in the area and said that Ball was "an Englishman who had adopted Normandy and whom Normandy has adopted". The works had waterwheels and steam engines rated at 235 hp and made both very heavy and fine papers. The raw materials included waste leather, bark, straw and waste fishing nets, but there was no mention of esparto grass. The production was 2000 tonnes per year and for 20 years the mill had worked day and night employing 150 workers at about 2.25 francs per day. The Balls provided paper for several government departments including the postal service and the war office. They also made paper for cigarettes, which became popular in western Europe after the Crimean war, where English and French soldiers had met the Turkish variety. However two years later in 1869 the Balls sold the Pont Audemer works back to *La Compagnie des Etablissements de la Risle*.[44] They moved to Sainte Addresse, a suburb of Le Havre, where Charles Ashby died in 1885 and Allen Charles in 1891.[44,59,60]

Papermaking continues to be a major industry of Pont Audemer but the original Ball factory site is now used largely for the storage of bales of waste paper. The buildings all appear to be 20th century but the arched channels which carried water from the former water-wheels are still clearly visible along the river. The old site is linked by a swingbridge, which carries a disused indus-trial railway across the Risle, to a modern paper mill and a cardboard box factory in the *Rue de Papetiers*. These are operated by *Papeteries Sibille Dalle* and *Cartonnerie de Pont Audemer* respectively.[56]

6. Conclusions

This paper is the result of a stimulating collaboration between a paper historian and a family historian and this method of research is recommended to all interested in the life and work of families of papermakers. It is particu-larly productive as marriage between families of papermakers was common so that a study of the history of one family soon provides information about others. The Ball family discussed here was active in both England and France and the research has therefore been both challenging and rewarding. In particular we have been dependent on the help of local historians and archivists in France who have no special interest in the paper industry or the Ball family but are enthusiastic about their own communities. We are conscious therefore that the research upon which this paper is based could have been carried out at greater technical depth and we hope that its publication will encourage French paper historians to collaborate with us in continuing our investigations. It would also be inter-esting to learn about other English papermaking families who moved to the Continent of Europe in the early 19th century. The Ball family certainly had good contacts as, in addition to making paper for the Comte d'Artois, Charles Ball II had a German mould-maker and exported paper to Russia. It is also interesting that in the late 1820s Hugh Rowland of Chilworth Mill, who succeeded Edmund Richard Ball at Postford Upper Mill in 1821, moved to a mill in Switzerland. Like Charles Ball III he made paper by hand in Surrey but set up a machine mill on the Continent and it would be inter-esting to discover how they gained the necessary expertise. The Ball family were also deeply involved in introducing straw as a raw material, a development which was mainly associated with France rather than England. The special interest of the Ball family however is that the paper industries which they helped to establish in France continue to this day in both Doullens and Pont Audemer.

Acknowledgements
We are indebted to the many archivists and local, family and industrial historians who have helped with the research upon which this paper is based. In particular we would like to thank David Barker (Albury), Bob and Retta Casbard (Albury and Postford), Robin Clarke (Surrey), Anne Knee (Merrow) and Tony Mercer (Albury and Postford) in England and Jean Avenel (Gueures and Val Vernier), Michel Laisne (Dieppe), Pierre Lefebvre (Doullens) and Irene Vautier (Pont Audemer) in France. Finally we would like to acknowledge Glenys Crocker, who has shared in much of the research on the paper mills, and Alan Bolton and other descendants of Charles Ball III, who have collabo-rated on the family history research.

References

1   Murdoch, T. (1985) *The Quiet Conquest. The Huguenots 1685 to 1985*. A H Jolly, London, pp.175-182.
2   Shorter, A.H. (1957) *Paper Mills and Paper Makers in England 1495-1800*. Paper Publications Society, Hilversum, p.221.
3   Hardingstone Parish Registers, Northampton: Charles Ball was buried on 17 November 1767.
4   Will of Charles Ball papermaker, dated 8 November 1767. Northamptonshire Record Office.
5   Farningham Parish Registers, Kent: Charles Ball was baptised on 17 July 1761.
6   Shorter, A.H. *op cit* in Ref 2, pp.179-200.
7   A special postal cover was issued in 1990 to commemorate the 150th anniversary of the production of this paper at Rush Mills. Penny Black stamps were on sale in London on 1 May 1840.
8   St. Clement Danes Parish Registers, London, Entry 498: marriage by licence of Charles Ball and Martha Weaver on 21 Sept. 1780. The licence, held at the Guildhall Library, London, ref 10,091/143, states that Charles Ball was 21 and Martha Weaver 19.
9   All Saints Parish Registers, Northampton: Charles Ball was baptised on 21 October 1783 and Edmund Richard Ball on 11 June 1786. The Charles Ball record gives the mother's name as Hannah and not Martha.
10 Shorter, A.H. *op cit* in Ref.2, pp.133-4. The index of persons in Shorter has about 1900 male names only seven of which are Edmund and two of these are Edmund Ball and Edmund Richard Ball.
11 Crocker, A.G. (1988) 'The Paper Mills of Surrey', *IPH Yearbook 7*, pp.10-17.
12 Crocker, A.G. (1988) *Paper Mills of the Tillingbourne*, Tabard, Oxshott.
13 Crocker, A.G. (1989) "The Paper Mills of Surrey', *Surrey History* **4**(1), pp.49-64.
14 Beauties, W.C. [c1800] Drawing of Stoke Mill. Minet Library, Lambeth LP 129/64/MIL.1.
15 Shorter, A.H. *op cit* in Ref.2, pp.167,227.
16 Crocker, A.G. (1984) 'Watermarks in Surrey Hand Made Paper', *Surrey History* **3**(1), pp.2-16.
17 Stoke Parish Registers, Surrey: Frederick Ball was baptised on 12 February 1792.
18 Tomb of members of the Ball and Collins family in St. John's churchyard Merrow, Surrey. Francis Wilcox Collins died on 3 January 1825 aged 6 months.
19 Drawing of Albury Paper Mill dated 13 July 1801. Northumberland (Albury Estate) Archives.
20 Maxted I. (1977) *The London Book Trades, 1775-1800*, Dawson, London.
21 *Gentleman's Magazine*, **90**(1), 1820, p.284.
22 *The Stationery and Fancy Trades Register*, 5 April 1869, p.196.
23 Master Papermakers' letter of 26 March 1796 on reverse of manuscript index to Russell's *History of Guildford*. Surrey County Library, Guildford, SP 942 21 GU1.
24 Albury Parish Registers, Surrey: Caroline Martha Ball was baptised on 29 April 1800.
25 Land Tax Records, Surrey Record Office, Kingston, SRO QS6/7 Merrow.
26 Merrow Parish Registers, Surrey: Edmund Richard Ball was married to Mary Anne Stanbrough on 13 January 1810 and Caroline Martha Ball to Francis William Collins on 2 November 1815.
27 Will of Charles Ball, late of Merrow but now of Clapham, dated 24 August 1818. Public Record Office, Chancery Lane, London, PROB11/1626.
28 Letter written by Frances Anne Raveau, daughter of Charles Ball III, in about 1887. Original held by Anne Phillips.
29 Letter dated 15 January 1819 from Ball & Ashby to William Henry Roberts, their representative in Scotland. Original held by Alan Crocker.
30 Anon. (1832) *Fourteen Views of the Village of Albury near Guildford drawn by a Lady*, Colagni & Son, London, p.11. Minet Library, Lambeth, 172.
31 Bank of England notebook (c1818) p.96. Copy provided to Anne Phillips by the Museum & Historical Research Section of the Bank on 21 April 1987.
32 *The Times*, 21 June 1820, no 10,965, 4e; 23 October 1820, no 11,071.
33 Sale Poster of Postford Upper Mill, 1820. Guildford Muniment Room 85/2/1(1) no 141.
34 Capel Parish Registers, Surrey: marriage of Charles Ball to Jane Elwood on 16 July 1806.
35 Portrait of Jane Ball (née Elwood), c1827-30. Original held by Anne Phillips.
36 Albury Parish Registers, Surrey: Charles Ashby Ball was baptised on 13 October 1809 (born 18 June 1809). Also St. Martha, Guildford St. Nicholas and Dorking Parish Registers, Surrey.
37 Guildford St. Mary Parish Registers, Surrey: baptism of Marion Martha Elizabeth Ball on 19 June 1812. St. Edmund Parish Registers, Lombard Street, London: baptisms of five children of Edmund and Mary Ball on 26 January 1822.
38 Dorking Parish Registers, Surrey: marriage of Jane Martha Ball of Dorking to Robert Thorley Bolton on 20 March 1828.
39 Documents on Paper Mill at Gueures. Archives Départementales de la Seine-Maritime, Rouen, 7 SP 485.
40 Bollengier, A. [1948] *Une Belle Figure de Gentilhomme Normand: Pierre Victor, Comte de Tocqueville*, Fournié, Paris.
41 Bilici, F. (1986) *Papier et Papeteries en Seine-Maritime (XVI-XX siècles)*, Archives Départementales de la Seine-Maritime, Rouen, pp.59-61.
42 Hunter, D. (1978) *Papermaking, the History and Technology of an Ancient Craft*, Dover, New York, p.536.
43 Moulin A. (1893) *Gueures. Annuaires de cinque Départements de la Normandie*, Caen et Rouen, pp.61-2.
44 Boase, F. (1965) *Modern English Biography*, **1**, Cass, London, p.145.
45 Information provided by Jean Avenal of Gueures.
46 Documents on Paper Mill at Val Vernier, Archives Départementales de la Seine-Maritime, Rouen, 7 SP 523.
47 Bégos, Y. (Mars 1989) *Brachy un Village autour de son Usine*, Connaissance de Dieppe et sa Région, Editions Bertout-Luneray, No.52.
48 Paris, Exposition Universelle (1855) *Catalogue Officiel Publié par Ordre de la Commune Impériale, Classe X, Section 5*, pp.67-8.
49 Portrait of Charles Ashby Ball. Original held by Anne Phillips.
50 Marriage Certificate of Charles Ashby Ball and Eugénie Herminie Camuset at East Malling, Kent, on 26 December 1839.
51 Register of Births, Doullens, 1833-42. Archives Départementales de la Somme, Amiens, 2E253/22.
52 Warme, M. (1863) *Histoire de Doullens*, pp.88-90, 147.
53 Cuerville, M. (1987) *La Memoire de Doullens*, p.286.
54 Information provided by Pierre Lefebvre, Mairie de Doullens.
55 Documents on Mills and Factories at Doullens. Archives Départementales de la Somme, Amiens, 251.
56 Information provided by Irene Vautier, Mairie de Pont Audemer.
57 Drawing of Pont Audemer Paper Mill in about 1850. Original held by Anne Phillips.
58 Shorter, A.H. (1971) *Paper Making in the British Isles*, David & Charles, Newton Abbot, pp.211,215.
59 Acte de Décès de Charles Ashby Ball, Mairie de Sainte Addresse, près du Havre, No.30, 27 Mars 1885.
60 *Journal du Havre*, 29/30 Mars 1885.

THE EVOLUTION AND DEVELOPMENT OF 'DRAWING PAPERS' AND THE EFFECT OF THIS DEVELOPMENT ON WATERCOLOUR ARTISTS, 1750-1850[1]

Peter Bower

For most of papermaking history papermakers had produced three basic types of paper: writings, printings and wrappings. 'Drawing' papers, which included papers for watercolours as well as those for pencil, chalk and ink, were quite simply those papers which artists found they could work on, regardless of the use for which the papermakers had actually designed and made them. During the eighteenth century however, under the impetus of various technological developments within the growing English paper industry and in response to outside influences such as the various wars with France which seriously interupted trade, makers began to produce very particular papers to satisfy some highly specific demands. Among these were 'drawing' papers.

From the 1750s onwards, a complex web of technical, cultural and economic influences operated on papermakers, paper merchants, artists' colourmen, and artists themselves, a process which gradually led to the evolution of papers designed and made specifically for watercolour. The effects of the introduction of such papers coupled with new sensibilities and aspirations in the artists themselves had a dramatic effect on the working practices of individual artists and their expression of their vision.

Watercolour, in particular, depends for its success on a deep and vivid understanding of the nature of particular surfaces and the ways that those surfaces can be worked. The ability of particular papers to satisfy the changing demands that artists now began to make on them was much influenced by the increasingly sophisticated technology of papermaking.

Following earlier technical developments in beating, mouldmaking and finishing techniques during the eighteenth century, the rapid introduction and exploitation of the paper machine, by an increasing number of papermakers in the first half of the nineteenth century, had the effect of making those who continued to make by hand change both their working practices and their concepts of papermaking itself, forcing them to produce papers of increased quality and range. They explored new raw materials and production methods, developed new qualities and characteristics; produced new surfaces, sizings, strengths, tones and colours. Watercolour artists seized on these new developments, not always using the new papers in the ways the makers expected, but grateful for the opportunities they provided.

Much of this increasing specialisation was due to the activities of merchants and colourmen who were differentiating more and more subtle developments in paper qualities and characteristics as selling points when marketing individual papers, both to amateur and professional customers. Artists themselves contributed to this process, telling the colourmen what they wanted. The evolution of medium-specific papers depended on a rapidly growing understanding amongst makers of the possibilities inherent in their process and an equal growth of understanding amongst users of their own needs and requirements.

Many of these 'drawing' papers were primarily designed for technical drawing, rather than for the fine artist. The market for fine artists papers has always been relatively small, but by the early years of the nineteenth century the market for engineering, architectural, surveying, and map making was enormous. It was this market that made the development of such papers economically viable for the mills.

One of the problems with the creation of increasingly specialised papers is that the individual sheets are in fact much less versatile, artists are less able to work specific papers to their own ends, than the papers designed for more general use. It is no accident that the advent of a generation of 'greats', Turner, Girtin and Cotman, etc, coincides with a period of papermaking when very specific drawing papers began to be made, but before such papers had become so specialised as to lose much of their potential versatility. A slightly later generation, in particular Cox and Palmer, were often to move away from these increasingly use-specific papers, both in their different ways becoming increasingly particular about their papers, and in the case of Palmer, later in his career getting London Boards put together for him by Newman, but then using extraordinary preparations to get the surfaces to his required state before working. Cox's two favourite papers were not actually drawing papers — one was a pure linen map paper and the other a hemp/linen wrapping.

Some artists such as J Duffield Harding put his name to drawing papers, which were made to his specifications, having a particular sale to the amateur market. Others such as Turner, De Wint, Varley and Cox were all to have their names appended to different drawing papers by both paper mills and artists colourmen during the nineteenth and twentieth centuries.[2] Perhaps the most successful papers, not necessarily in commercial terms, but in use, in the popularity of them with practising professional artists were two specific artists papers developed during this period: the paper called *Creswick* [3] and the various toned and coloured papers made by Bally Ellen and Steart in the 1820s and thirties. Both are essentially simple papers: pure linen, gelatine sized, made with great care and attention. The range of papers used by artists is as varied as the artists themselves and their ways of working. Part of the fascination of an on-going study such as this is the ways in which different artists have chosen their papers, and the range of papers they work on.

The production of papers specifically for drawing began sometime in the 1770s but was not to become a major part of the handmade paper industry until the early years of the nineteenth century. Continuing analysis of actual papers used by artists, and research in various archives is slowly pushing this date back further into the 18th century. The earliest examples of the increased specialisation in artists' papers for were the development of fine printing papers for engravings, necessary to fill the gap left by the disruption of the paper supply from France, caused by various periods of war with France and related trade embargoes. Two fine examples of works which illustrate some of the dramatic changes that took place in both papermaking and artistic practice, during the period under consideration, are Thomas Gainsborough's (1727-1788) 1785 drawing, *Figures in a wooded landscape* [4] and John Frederick Lewis' (1805-1876) 1853 watercolour, *The Noonday Halt* [5].

The contrast between the two papers used for these works is striking. Gainsborough's choice was a pale laid *papier bulle*, made from a blend of low grade linen and old hemp rope, with a slightly coarse surface and a very prominent laid impression from the mould[6]. This paper type, originally designed as strong wrapping paper, was manufactured by a very large number of small provincial mills throughout western Europe. It had been popular with artists over the preceding two centuries, for the subtlety of its warm off-white to buff colours that provided them with a perfect mid-tone, particularly useful for chalk drawing, but also used with wash and colour as well.

In total contrast, the smooth finished, very fine quality pure linen paper used by Lewis, for his painstakingly detailed watercolour, had been made with precisely Lewis' working methods in mind. It was capable of taking paint in ways unthought of in Gainsborough's time.

The seventy years between those two works saw revolutions in both papermaking technology and practice and in the needs and aspirations of those artists working in watercolour. Between the dates of these two works everything in papermaking changed: raw materials and their preparation, the design and construction of the moulds used in forming the sheets by hand, presses, sizing and finishing techniques and drying methods. By the time Lewis painted this beautiful and startling image, the colourmen were selling an increasingly large range of very different papers, in some cases up to 80-90 different papers, for the most part designed for very specific areas of the fine art market.

These two images are a good example of the outcome of the very complex set of influences that had operated in the years between the making of the two works, involving artists, colourmen, merchants, stationers and paper-

Figure 1: *Bere Mill, Hampshire, typical of the corn mills converted into paper mills early in the 18th century.*

Figure 2: *Frogmore Mill, Hertfordshire, converted from handmade to machinemade paper-making in 1803.*

makers, to develop such media-specific papers. When discussing the market for artists materials it is worth distinguishing between the amateur and professional markets. Much of the paper designed for artists in the nineteenth century was designed to appeal to the amateur, the professionals were less likely to be taken in by the often very dubious claims of the colourmen regarding particular papers. The professionals also seemed more inclined to experiment with very different papers, not designed for watercolour, but often capable of taking considerable working.

We will see, however, that the increasing emphasis on highly specific characteristics in the design of 'drawing' papers inexorably led to a sacrifice of versatility in those papers. This led to some artists, one thinks of Cotman, Turner, and Cox in particular, to react against some of these developments, moving away from these new papers, searching out other papers, often designed for very different usage, which would give them more range of possibilities in the way they made their marks. Other artists, such as Palmer, were, at various stages in their careers, to evolve quite complex ways of preparing their papers that again allowed their individual talents full rein.

In general, English paper usage for watercolours, before the 1790s, had been based on the use of fine writing papers. Although designed for use with ink and a quill, and later the steel nib, these often had the necessary highly developed surface strengths suitable for water-colour washes. Paul Sandby's work (1731-1809) shows many fine examples, both in his finished works and in more informal sketches of such usage[7]. These papers had the necessary surface strengths and light reflecting capabilities, unlike printing and plate papers which were too soft-sized. Many wrappings were also just not strong enough, though as we have seen with the Gainsborough example discussed above many buff, blue and grey wrapping papers were used.

Artists have always chosen papers that suit their particular needs. If they couldn't find what they wanted then they would alter the sheet, as is seen in the work of Alexander Cozens (c1717-1786). The yellow prepared grounds he favoured were, in all probability, applied by Cozens himself. It was possible to buy such toned and coloured grounds prepared by artists' colourmen, but there is no evidence of these yellow tones being available commercially at this date, and Cozens' use of such grounds is so individual it suggests he was preparing them himself[8].

The slow evolution of "drawing papers" owes a lot to the idiosyncrasies of individual artists' working practices. It was difficult for the mills, and the colourmen, to know quite what was desirable. If we contrast Francis Towne's (1740-1816) 1781 watercolour *The Source of the Arveiron*[9]

and Thomas Girtin's (1775-1802) 1801 watercolour *Ilkley, Yorkshire*[10], we can see two very different approaches. The Towne is on a high quality writing paper and the Girtin on cartridge paper. Cartridge paper, as its name implies, was originally developed for wrapping powder and shot, but its hard sized strength and rugged surface soon led to it being used for other purposes, as a wrapping paper as well as for drawing and painting. Artists were attracted to both its strength and versatility and its warm off-white or pale buff tones. Although Girtin is renowned for his use of Cartridge, a very high proportion of his works are not executed on cartridge paper as such, but rather on pale whited-brown wrappings where some bleaching has shown up the ligneous parts of the low grade linen and hemp fibres used in the making of the sheets.

These two works illustrate the contrast between the superimposition of a vision onto the sheet of paper and the almost instinctive understanding of how a particular vision can be realised. Towne has forced his image onto four individual pieces of paper carefully laid down. Watercolours by Towne, which are executed on two or four joined pieces, have often been described as being on pages removed from a sketchbook, but, with two or three exceptions, all the examples I have seen have been on trimmed sheets rather than sketchbook pages. It seems impossible to work out what he was trying to do by this joining together of smaller sheets when bigger sheets were available. If this had occurred merely once one might presume that he wanted to use this particular paper and that all he had were some small sheets, but this occurs over and over again in his work, with this and other papers. In the Girtin the paper plays its own part: it is very much part of the image, rather than merely a ground to carry that image.

Francis Towne's friend and pupil, John White Abbott (1763-1851) gives us some even more extraordinary examples of an artist joining many small pieces of paper together. In one small work, only 431 x 355 mm, *Trees in Peamore Park Exeter*[11] he uses no less than nine separate small pieces of the same white laid fine writing paper, assembled in a complex mosaic, with no regard to placing the felt or wire sides together to produce a coherent surface across the whole of his assembled working surface, or indeed with the fall of the laid and chain lines.

We have looked at examples of writing papers used in the 18th and early 19th centuries for watercolour. Very few printing papers were used because they were generally too soft-sized for use with washes or water-colour, but one other type of paper was used a great deal, and has been popular with artists for centuries: wrapping papers. These were made in a range of colours, where the colour came from the raw material used rather than from a dye added to a basic white pulp. They came in a range of sizes, with lovely names, *Double Loaf,*

Wedding Purple, etc, and in a range of weights and sizings. A range of very popular papers were the so-called 'academy' papers, one brown and one blue, used by many artists including Turner in his days as a student at the Royal Academy Schools. These strong wrappings, originally chosen by the colourmen as suitable for chalk drawing, came to be called Academy papers because of their use in the Royal Academy Schools. By the early years of the 19th century the colourmen were having coloured and toned papers made for them to replicate these papers "in a great variety of tints and finishes" with "numbered sample books available on request"[12].

One artist who, with Turner and perhaps Cox, had an extraordinary understanding of surface strength and texture, of knowing what the paper will give him was John Sell Cotman (1782-1842). His *Bedlam Furnace*[13] is on a white wove, pure linen, hard sized, drawing paper of the finest quality, possibly made by Edmeads and Pine, the surface is rubbed and worked almost, but not quite, to extinction, just far enough so that the beginnings of its surface breakup actually contribute dramatically to the vivid quality of the light that he wants to capture in this image. In complete contrast, *On The Greta*[14] has been executed on a hemp-linen mix, much softer sized, buff laid wrapping, carefully worked and well understood, the very colour of the sheet suffusing every wash and mark he made. In contrast to the vigorous scrubbing and the speed of some of the marks in *Bedlam*, Cotman would have to have worked relatively slowly to produce the very distinctly drawn areas of colour in this Greta drawing.

Comparing two artists' usage of similar papers to achieve similar ends can be instructive: Turner's *Storm off the East Coast*[15] and Cox's *A Train near the Coast*[16] use very similar techniques on visually similar papers but those similarities disguise some very real differences. Cox's paper was one of his 'scotch wrappings', and although the tone of the paper used here by Turner is very similar to the Cox paper, this work by Turner is on a completely different paper, which was made as an artists' paper, probably by Bally Ellen & Steart of De Montalt Mill, near Bath. Visually these buff, brown and indeed the blue papers, made by Bally Ellen & Steart as artists' papers, owe a lot to eighteenth century wrapping papers, but with the best papermaking developments of the 1790s-1820 added in.

It is also instructive to examine an artist's use of different papers but for the same image; De Wint's sketch for *Torksey Castle*[17] and his finished work[18] are designed for two different purposes and executed on very different papers. His swift study of this subject had been executed on a Rough surfaced "Creswick"[19], admirably suited to his confident brush. The much larger, finished, "exhibition" piece has been executed on a Hot Pressed Whatman with a heavy degree of gelatine sizing, more suitable for the layered and almost luminous washes De Wint felt appropriate to the finished view of this particular subject.

The things that artists were actually doing on and with particular papers gradually filtered back to the papermakers via their agents, the colourmen and merchants, leading to some of the changes in papermaking practice that would bring about such dramatic differences in the capabilities of individual papers. The mills themselves changed throughout this period, Bere Mill (Figure 1) was built on the upper Test in Hampshire by 1710 as a flour mill, but in 1712 was purchased by Henri Portal and converted to papermaking. The attic spaces of the old mill still contain the wooden chutes designed to supply the millstones with corn from the storage bins. The photograph of Frogmore Mill, (Figure 2), was taken about 1880 and shows a handmade mill adapted for machine usage. Frogmore was the mill where the world's first production machine was up and running by 1804. The basic dynamic of papermaking did not change, merely the scale and detail of the operation.

One of the most important and visually obvious changes, and one of great interest to artists, was the development, in the mid 1750s of wove paper. Despite the visual importance of the distinct differences between wove and laid paper, this distinction has little direct relevance to the less visible, and more important, changes that were going to take place, which all actually effected surface strengths of the sheet, something crucial to watercolour artists, and the consequent changes in the behaviour of individual papers in use. As can be seen from figure 3, all laid papers, even if hot pressed, carry shadows in their surfaces, which can have a profound effect on the way any pigment, any mark made by the artist, reflects light back at the viewer. Such shadows can change the tone and effect of any colour, particularly towards the red end of the spectrum, because the subtle shadows in the texture of the surface are closer to the blue end of the spectrum anyway.

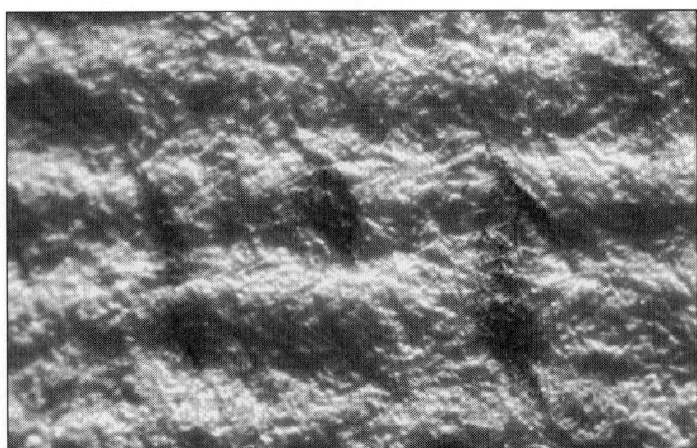

Figure 3: *Close up of a laid paper showing the shadows still prominent, even in a smooth sheet. Contrast this with the wove surface seen in Figure 4.* (Photograph by Marcus Leith)

Gainsborough was said to have offered one of his "Landskips" for a quire of the new paper, so keen was he for this smooth and different surface, to Robert Dodsley, the publisher of *The New Bath Guide*, who had used some wove paper for the fifth edition of the book. He wrote that

> There is so little impression of the Wires, and those so very fine, that the surface is like vellum...[20]

Gainsborough was sadly disappointed when Dodsley sent him some fine laid paper, not the wove:

> I had set my Heart on getting some of it, as it is so compleatly what I have long been in search of: the mischief of that you were so kind to inclose, is not only the small wires, but a large cross wire... which the other had none of, nor hardly any of the impression of the smallest Wire... I am this moment viewing the difference of that you sent and the Bath Guide holding them edgeways to the light, and could cry my eyes out to see those furrows.[21]

The early woves would have been well suited to Gainsborough's chalk drawings, but given that they were actually made for printing, most were at this date much too soft for any vigorous watercolour work. That Gainsborough was eventually able to acquire at least one sheet of wove paper can be seen in his late great presentation drawing, *A Mountainous Landscape with a Herdsman and his Cattle passing a Cottage*, the only known Gainsborough on white wove paper. His happiness in working on such a surface is immediately obvious to the eye, it is a work of great spontaneity and simple boldness.[22]

The 1700s had seen the rapid development in England of a home grown 'fine' paper industry, based on a new technology imported from Holland and the more tradi-

Figure 4: *Close up of a wove sheet, showing the much more even spread of surface shadow than is possible in a laid paper.* (Photograph by Marcus Leith)

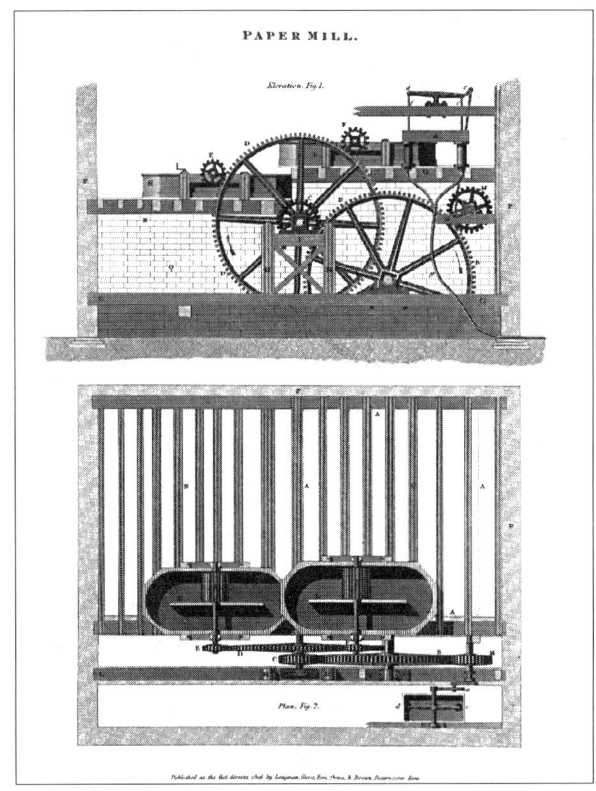

Figure 5: *An 1816 breaker – beater combination, from Rees' Cyclopaedia, shows increasing sophistication in the design and use of papermaking equipment.*

tional skills of the Huguenot refugees from France.[23] An added impetus to change were the various wars with France throughout the Century which stopped much French Paper, previously a mainstay of English paper consumption, from entering the country. The combination of the introduction of wove paper with newly developed techniques for the choice and preparation of fibres was to have a decisive effect on what artists could actually achieve on the surfaces of their papers.

An old saying in the paper industry is that "Paper is made in the beater". This means that it does not matter how good you are at all the other stages of the papermaking process: unless you get the choice and preparation of the fibre right the paper will simply not do what is wanted of it during use. The stampers used for pulping, although capable of producing very fine pulps, did not allow great subtlety in the beating process. The advent of the hollander beater allowed papermakers a different understanding of the sophisticated possibilities inherent in particular fibres, and to explore their potential in this new machine. They began to appreciate that the degree of beating, the wetness or freeness of the fibre, the length and speed of the beat, the weight of the roll, the angle and configuration of the bedplate and the roll bars could all contribute to the creation of very different

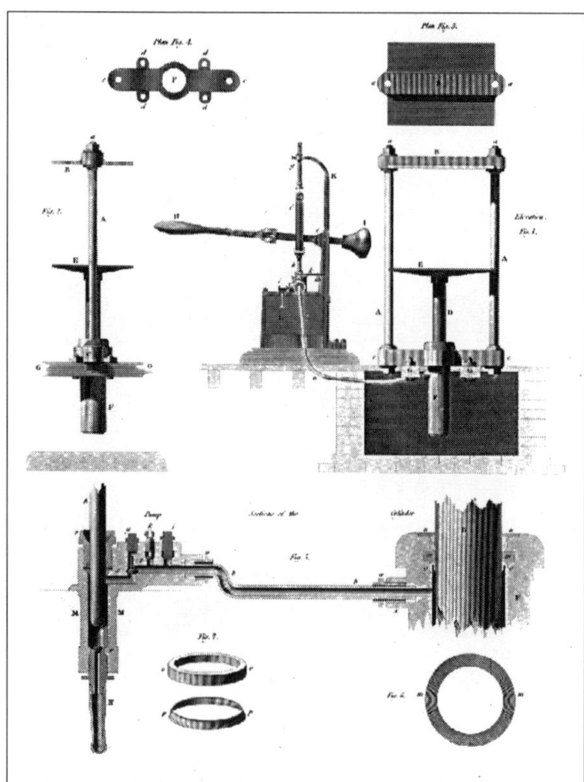

Figure 6: *Joseph Bramah's "Hydrostatic Press",
developed between 1795 and 1805, speeded
up production and effected the kinds of
finish available in the pressing of wet paper.*

behaviour in the finished sheet of paper, even when the base raw material was the same.

The growing understanding of the basic pulping process led to an increased awareness of the smaller and smaller distinctions between types and grades of raw materials, particularly in the sorting of rags. In the 1780s there were perhaps 8 kinds of easily distinguishable rags, by the 1850s over 100 distinct categories were on offer from the rag merchants. The blending of fibre, either of different fibres, or different qualities of the same fibre has occurred throughout papermaking history. Consistent supplies of rag fibre have always been a problem. What had generally been done for reasons of cost, sometimes aesthetics, and often forced on the papermaker by short-falls in particular supplies, now became more scientific, as the characteristics of different fibres could be developed in the beater to achieve desired ends in the behaviour of the finished sheet in use. Many paper-makers took up such challenges with more or less success. They began to produce much more flexible sheets, more versatile papers, that were more capable of taking the increasingly demanding work of the water-colourists.

Many of the changes that were taking place were 'invisible'. Quite simply, technical innovations and the resultant new skills drove papermakers, in their increasing specialisation, to produce more and better papers, and produced some profound changes in the ways sheets behaved in use, but many of these changes were not always visible. The development of new types of presses, primarily to speed up production and to give greater strength for writing papers (which had had to change with the introduction of the steel nib, which demanded a much greater surface strength than the quill) also gave greater pressure and compacted the surface of the sheet differently. Improvements in felt manufacture were also occurring at the same time and some makers began to develop new and more complex wet pressing sequences: not just one quick press to remove the water, but another pressing with dry felts, another without felts and with less pressure, another with all the sheets in a different order, etc.

There is evidence that Koops had installed, or at least wanted to install a hydraulic press at Neckinger Mill in 1800, some years before Bramah actually took out his patent for such a press[24]. The changes in these pressing techniques had a considerable effect on the actual surface strength of the sheet, as did the introduction of larger amounts of cotton rag into paper making. Linen has the characteristic when being wet-pressed of having a much higher relative surface strength when it is a thin sheet than when it is thicker. Some of the lightweight writings that Turner used for many watercolours show just this kind of characteristic. A cotton rag or cotton linter based sheet of the same weight and bulk would quite literally not be able to take the kind of vigorous and often many layered working that he subjected them to. Cotman is another artist who, in his multilayered build up of washes took advantage of this characteristic. This increase in surface strength, and the differences in the way the sheets were pressed also had an effect on the way size worked on the surface of the sheet, and changed the absorption of the surface, allowing coloured pigment to sit on the surface differently and more brightly.

Sizing itself went through many changes. The older heated tub where a few sheets at a time were sized eventually evolved, via various intermediate forms and techniques, such as Whatman's box techniques described by Joshua Gilpin[25], into the type of size bath seen in Figure 7.

The size itself changed, from gelatine to rosin, but most of the handmills continued to use gelatine for the finest papers until well into the middle years of this century. Perhaps a more significant change however was what the gelatine size was made of. The use of *scrolls*, the parings from the vellum and parchment makers, or the use of kid leather for plate papers[26] and other highly specific methods gave way in many mills to a general use of rabbit skin, fish, bone and by the mid nineteenth century in England at any rate, the import and use of buffalo hide from the United States. All these sources have distinct effects on both the appearance and the behaviour of the sheet. Scrolls gave a very clear gelatine, hardly changing the actual colour of the fibre at all, whilst rabbit and more particularly buffalo were very 'yellow' changing the tone of a white sheet quite considerably.

The cleanliness of the size house and the utensils used as well as the length of time a batch of size was used also had a great bearing on the quality of the finished sheet. Some mills were punctilious about their size, others used the same batch over and over again, filtering and reheating it until it really ceased to have any life. By this stage it was also usually rather dirty, giving a dull cast to the sheet.

Any gelatine-sized handmade sheet goes through two drying processes, once just after it is formed and then again after it has been sized. The first drying is crucial to both the stability and strength of the sheet. All paper, but particularly watercolour paper, is best dried over a long period using airflow and gravity. Speeding up the drying process particularly by the use of heat, has the effect of not allowing all the tensions within the paper to resolve themselves right out to the edges of the sheet. Each sheet is subject to enormous internal tensions as it dries, shrinking as it gives up its water to the air around it. Each fibre gives up water and as it does so it tugs on the fibres with which it is mechanically entangled, creating waves and ripples, visible in the surface of the sheet. This process is known as cockling and is more or less visible in most handmade papers.

Drying a watercolour sheet in the winter gives a much longer length of time for these tensions to resolve right out. It is no accident that the papers that allowed English water colour artists to develop their art in new and perhaps surprising ways were made in a climate such as ours. French papermaking, particularly makers such as Columbier, Dupuy and Richard from the Auvergne, was renowned in the eighteenth century for the extraordinary bite and definition that could be achieved with engraved and etched plates. Papers dry much faster in the Auvergne, giving less time for shrinkage and thus producing a slightly 'softer', less dense, sheet that allowed the engraving plate a beautiful bite into the surface of the sheet. Because these plates were rarely run through the press more than once there was no intrinsic reason why the papers needed great internal stability. The difficulty many English makers had in matching the quality of French plate papers in the 18th century is partly due to the difference in these drying times.[27] Heat began to be used by some makers in the late 18th century to speed the process up, but the first drying was still usually undertaken without heat.

A major change in drying did take place however. The traditional method of drying over cow or horse hair covered ropes (ropes that unlike hemp would not mark or discolour the wet sheet) had always had one disad-

Figure 7: *Gelatine sizing paper Tuckenhay Mill, Devon in 1900.*

vantage: when first dried every sheet had what was called a *back*, a bend in it where it had hung over the rope. This area of the sheet was often distorted by differential shrinkage leaving creases and cracks in the surface as well as distorting the actual shape of the sheet (see Figure 9). A second result is the distortion of the long edges of many handmade sheets, dried by this method, which are slightly curved, where the middle of the sheet has not shrunk as much as the outer edges. For many uses this did not matter too much, paper was generally sold folded in quires of 24 sheets and the *back* was

Figure 8: *Drying over ropes at Springfield Mill, Kent.*

Figure 9: *Close-up of the ripples in the surface sheet paper caused by differential shrinkage when the sheet was dried by being hung over a rope.* (Photograph by Marcus Leith)

hidden in the fold. Much paper was sold for printing where again it would not matter[28].

Many artists didn't mind the presence of the *back*, there are many examples including several famous paintings by Turner, Cotman, Cozens, Joseph Wright of Derby, and Paul Sandby, where the back is clearly visible in the work and has sometimes, sadly, been 'read' as damage by some conservators who desperately tried to do something about it much to the detriment of the work.

But some artists did not like it, particularly as they desired to work larger and larger. Colourmen such as Ackermann were also convinced that the paper would be better without it and when a method of drying that did away with the back was developed they marketed it extensively. The merchants were always looking for new papers to present to their customers, whether amateur or professional, and many artists used such highly developed papers for some of their works, whilst continuing to find other papers to use as well. *Seamless drawing papers*, ie papers without this disfiguring line were being marketed by Winsor & Newton and others quite early in the 19th century.

Seamless drying, where wet sheets are laid flat in spurs, on muslin or coarse canvas 'sails', was not a new process. Any information as to when this practice actually began would be warmly welcomed. There are many large watercolour papers from the earliest years of the nineteenth century which show no trace of a rope mark or "seam" anywhere on them, but the earliest reference I have yet found to this method of drying, particularly being used as a selling point in relation to watercolour papers, comes in a 1861 trade catalogue issued by Winsor and Newton where, after the entries for "Whatman's Drawing Papers" and "Whatman's Drawing Paper of Extra Weights" can be found "Whatman's Seamless Drawing Papers". The note above the sizes and prices reads:

These Papers are similar to the foregoing stout Drawing Papers, but are perfectly flat, and without any seam mark across the centre of the sheet.[29]

The same Catalogue also advertises another paper, dried using the same method: Winsor and Newton's *Griffin Antiquarian*, specially made for Winsor and Newton by Balston at Springfield Mill. It is stated that

great advantage will be found to result from the Antiquarian having been dried on the new flat principle, where *the line mark across the centre is avoided, and a uniform surface presented.*[30]

The works of two very different artists, Samuel Palmer and David Cox, at different times in their careers, show the importance that differences in vision, temperament, ways of working, habits and aspirations, have in determining the choice of paper an artist will work on. Many artists could have been chosen to illustrate such a point, but Palmer and Cox have such significant differences, and some intriguing similarities that contrasting their work allows some very important points to be raised. I have given some further details of the paper usage of three other very individual artists from the same period, William Blake, John Constable and John Sell Cotman, in the Appendix.

As with many artists Palmer worked on a variety of papers throughout his life, many of them also worked on by other artists, but for his late series of very complex watercolours, based on poems by Milton, he developed an extraordinary system of preparing his working surfaces. A letter from Samuel Palmer to Richard Redgrave describes some of his curious methods and is worth quoting from at length[31]:

...about the board I work upon. It is a sixsheet "London Board" of best quality from Newman's, Soho Square[32] (which I think a most reliable place for watercolours), Imperial or Royal: the royal, they say, is made for or used by flower painters.

I take off the gloss by rapidly sponging with water (a *little* alum dissolved in it might perhaps do no harm) rapidly lest the paste should soften underneath; thoroughly, so far as to see that it takes the water everywhere. I was driven to use it simply because, on the best modern paper (the only sort to be had since Creswick's was bought up[33]), I found it impossible to get any one quality I liked.

At first I felt like a baby on a slide, but found that this might be avoided by using the

Bower: Development of 'Drawing Papers'

colours less wet than on a simple paper. Ultramarine especially, and the ashes, always difficult to wash, will not bear a full, wet brush; but a very thin wash of zinc white over the sky and distance part, before beginning, might facilitate. Several practised artists have done this.

These difficulties have their equivalent advantage in the ease with which a bit of sponge or even a wet brush will remove dried colour. In this respect the surface is intermediate between paper and ivory[34]. I have done my largest things upon it; getting it made on purpose rather than use paper[35]. I fasten it around the edge with the smallest flat-headed brass pins (not passing them through it, which if you avoid working on parts already wet, prevents warping), upon a thin, light, drawing-board of seasoned pine... a sheet of white paper is fastened on the front of the drawing-board, so that even if you have not much margin of the "London Board" you have a sort of white mount...[36]

According to Palmer, a London Board, prepared in this manner

> reflects light through the pigments. It shows the full depth of the colours laid on, and as you retouch with more colour the retouchings are still darker, as one would suppose they naturally must be. In the modern papers, you come to a full stop at a point far short of the depth of the pigment — in fact it is an ocular paradox.

Palmer had very distinct views on the right kinds of papers depending upon what sort of work was being contemplated. Two further extracts from other letters show his particular fondness for creamy whites for painting and etchings and brownish-grey tinted papers for sketching

> Dark pictures are best drawn on brown crayon-paper (the tint of Whitey-Brown, but a little darker)[37]

> Water colours upon paper hued like the lightest whitey-brown paper (I have a sketchbook of whitey-brown paper which Mr Newman got hot-pressed for me)[38]

Palmer hated some of the developments in artists' papers that the merchants, and the papermakers were insisting on

> Supposing there should be a lucid interval in this vile grey and blue paper system... it is

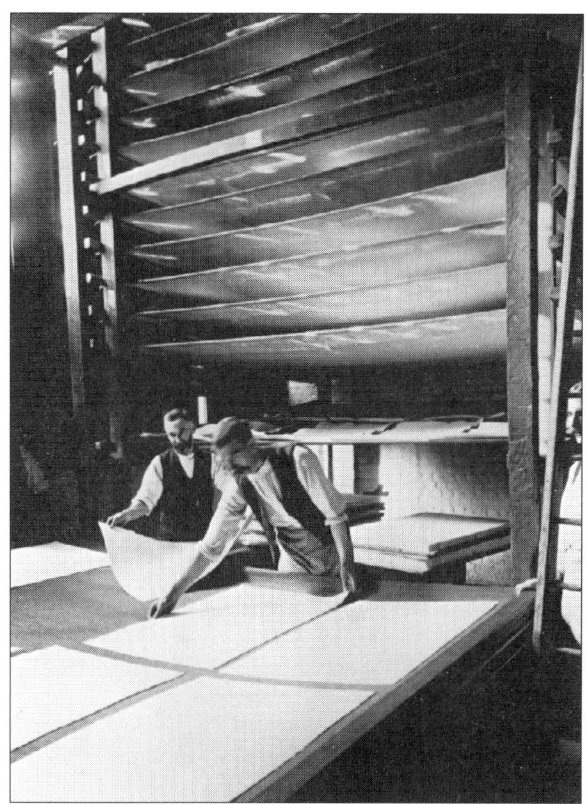

Figure 10:*Sail drying at Springfield Mill, Kent.*

> evident to me that something has been done either to blue or bleach the paper. If we asked the papermaker to make us a creamy tone we should rue the day. All I say is, after you have washed them, make up the rags, pure and simple or impure and simple, into paper, just as they are, and let them alone: no bleaching acid, please, no blue bag, no toning... My demurs as to papermaking are a trifle. I was told at Mr Colnaghi's that a chemical agent was used in making the paper for a large and expensive engraving, which after a good while turned it BLACK.[39]

Both the colourman, Newman, mentioned by Palmer, and the papermaker and pasteboard Creswick were only two of many such makers, who specialised in the manufacture of two, three, four, six, or even eight-ply laminated drawing boards. The biggest suppliers of such 'London', 'Bristol' and 'Drawing' boards to the London market were Turnbulls, of Holywell Mount, London[40]. Their products can be found in use by artists all over Britain and are listed in many of the trade catalogues issued by the London colourmen such as Reeves, and Winsor & Newton[41]. They are generally stamped with a range of blind-embossed stamps, either vertical or horizontal ovals, bearing the Arms of the City of London and with various wordings.[42]

It has often been supposed that the distinction between 'London Boards' and 'Bristol Boards', most of which were made by the same manufacturers, lay in the choice of papers: that 'London Boards' were made only from the finest Whatman paper and that 'Bristol Boards' may have had one ply, the intended working surface, utilising Whatman paper but that the other sheet(s) would be made-up using papers of a lesser quality. This may well have been the case in the earliest years of the nineteenth century, but recent research in the archives of the Royal Society of Arts has shown various examples of such boards that go against this interpretation, being made up of papers by very different makers[43].

Samuel Palmer and David Cox were two men who could not have been more different in their ways of painting. Cox, like Palmer, had very specific preferences when it came to his materials, particularly the tones, absorbency and strengths of his working surfaces. He liked papers that allowed him to work fast with a heavily loaded brush. Before his discovery of the paper which will forever be associated with his name, the famous "Scotch Wrapping", he had two papers that he really liked. One was what he called his 'old favourite cartridge'[44] which he continued to use throughout his life and the source of which has never been satisfactorily identified. The second was a paper which he acquired from his brother-in-law, Gardener, who:

> had the sale of the Government Ordnance Maps (and) was commissioned to get some superior paper made for them at the paper mills at Uxbridge. Some sheets were made beyond the order, and these and some cuttings of the paper were obtained for Cox, who prized them as being superior...[45]

In 1836 Cox had found what he, from then on, called "Scotch Wrapping". It was made from bleached sailcloth and old hemp rope and designed for wrapping reams of better quality paper. He had obtained a few sheets by chance from Grosvenor Chater[46] and was told by them that the Excise Mark 84B indicated that it was manufactured at Dundee[47]. Impulsively he ordered a ream which to his amazement weighed 280lbs and cost eleven pounds. Cox couldn't actually afford the eleven pounds but was helped in paying for it by an old friend, the amateur watercolourist William Roberts, who also had to help Cox carry the ream away from Grosvenor Chater's office. Cox's initial dismay at the cost of the paper soon turned to regret that he did not order more, as the paper exactly suited his needs, and he could never find any more of equal quality. Its ready absorption of colour and the particular nature of the rough texture were admirably suited to his vigorous strokes made with a full wet brush. The paper also contained many small black and brown flecks[48]. On one occasion when asked what he did with these specks if they fell awkwardly, say in the sky,

he is reputed to have replied "Oh, I just put wings on them and they fly away."[49] A particularly good example of his use of this paper is his picture *Sun, Wind and Rain* of 1845[50].

Watercolour, whether a sketch or a full-blown exhibition piece, is an essentially intimate work, where the surface and the paint on it have to be seen close to. The actual interplay of the surface and the marks made on it are an integral, often crucial part of the actual visual effect of the work, something understood by those artists who excelled in this deceptively simple medium. Many artists have left no real evidence of their thoughts on the various media they used, save for the actual works themselves, which are perhaps the most eloquent testimony we could have. Turner probably summed up the importance of paper to watercolour best, when in response to a request for advice on watercolour painting he replied "First of all, respect your paper!"[51]

Notes and References

1. This paper is essentially an introduction to a very complex area of the relationships between papermaking and artistic practice, and which forms the basis of a book currently being written. An earlier version of this paper was first given at *The Great Age of British Watercolours* Symposium at the Royal Academy of Arts, London, March 1993. When this paper was given at the BAPH Conference it was mainly illustrated with coloured transparencies of individual works of art. The constraints of this publication have meant that rather than use these works to illustrate this paper I have concentrated on using those images used during the presentation of the paper which do not depend on colour for their success. For those interested I have given, in these references, the museums, galleries and institutions where the individual works discussed can be seen.
2. *J D Harding's Pure Drawing Paper* was being marketed by Winsor and Newton as early as 1845. *Turner Grey* and *De Wint* were both made by J Green & Son, and later Barcham Green, at Hayle Mill, Kent. *Varley* was made for the colourman James Newman, whose publicity in 1910 was claiming that it had "originally been made by James Newman for that celebrated artist John Varley". *David Cox* paper was a very popular machinemade flecked buff paper made for Dixon's, the paper merchant, based, very loosely on Cox's favourite Scotch wrapping. Production of Dixon's David Cox paper ceased in the 1960s.
3. After Creswick stopped manufacture another version of this paper was made by Balston at Springfield Mill and marketed as "Imitation Creswick".
4. Private Collection (Leger Galleries, London).
5. Fitzwilliam Museum, Cambridge, 716.
6. Unfortunately the Gainsborough drawing has been rather unsympathetically conserved at some point in its history. The surface is now very flat, but the original deep laid impression is easily visible in Gainsborough's rich chalk drawing.
7. eg: Paul Sandby, *Bayswater*, 1793, British Museum, London, 1904-8-19-67.
8. eg: Alexander Cozens, *Figures by a Pool*, c1765, Hazlitt Gooden & Fox, London.
9. Victoria and Albert Museum, London, P. 20-1921.
10. Leeds City Art Gallery, 5. 113/92.
11. Fitzwilliam Museum, Cambridge, PD 42-1980. Another example of the use of very small cut down pieces of paper

is *The Townsend Album* at the National Portrait Gallery, where an extraordinarily high proportion of the very small drawings contain watermarks. One wonders what happened to the rest of the sheets if so many of the centres of half sheets of writing paper have been used for these drawings.

12. Trade Catalogues from various dates, Winsor & Newton Archives, Harrow.

13. Private Collection.

14. Leeds City Art Galleries, 16. 2/55.

15. Sheffield City Art Galleries, 2205.

16. National Museum of Wales, Cardiff, 3036.

17. Usher Art Gallery, Lincoln, U.G.71/88.

18. Usher Art Gallery, Lincoln, U.G.86/13.

19. This sketch has probably been overcleaned at some point in its history. The particular warm off-white to light buff tones of much "Creswick" when made is unfortunately rather reminiscent of the discolouring seen in so many 19th century papers. Because of this similarity they have all too often been bleached back to a whiteness that they never had originally, thus interfering with the integrity of the work. The artist chose to work on a particular tone and making the ground whiter distorts the original colour balances in the work.

20. John Hayes, *The Drawings of Thomas Gainsborough*, London 1971, Vol I, p 21.

21. Hayes, *ibid*.

22. This work was exhibited at the Leger Galleries, London, in 1995.

23. The introduction of this new technology into England is discussed at length in John Balston, *The Elder James Whatman*, 1992, Volume I.

24. Joseph Bramah, Patent No 2840, 25th April 1805.

25. The unpublished *Journals of Joshua Gilpin*.

26. As specified by James Basire for *Antiquarian* paper to be made by James Whatman at Turkey Mill, Maidstone.

27. see Brenda Weller, *Intaglio Printmaking papers and the RSA*, unpublished thesis.

28. The *back* would disappear into the spine (folio) or be trimmed off in the binding (top edge - quarto, long edge - octavo, top edge - 16mo, long edge - 32mo etc).

29. *Winsor & Newton Drawing Paper Stationers*, Catalogue for the Trade Only. Issued November 1st 1861, page 4. (Winsor & Newton Archives).

30. *ibid*, page 7. Although made by Balston this paper does not carry the Whatman watermark. Each sheet was watermarked W&N and also carried a blind-embossed stamp of a *Griffin* in one corner. The Artists' Colourman, James Newman, also had Balston make special 'Seamless' drawing papers for him. These can be identified by the large copperplate N next to the usual J WHATMAN / DATE in the watermark. Papers made by the Hollingworths at Turkey Mill for Newman also had the same copperplate N added to the watermark.

31. A. H. Palmer, *The Life and Letters of Samuel Palmer*, London 1892 (1972) Letter XLVII to Mr Richard Redgrave, July 14 1865.

32. James Newman was one of the oldest established Artists' Colourmen in London, specialising in very particular papers, many of which "Whatman" papers, made to his own design, either by Balston at Springfield Mill or the Hollingworths at Turkey Mill. See Footnote 30, above for details of Newman's distinctive Whatman watermarks. After the Hollingworths discontinued making by hand in 1859, one sees the Hollingworth J WHATMAN / TURKEY MILL mark with the Newman copperplate N and an additional large outline Roman letter B, presumably to indicate that despite the Turkey Mill in the mark, such sheets were actually made by Balston at Springfield.

33. Thomas Creswick was a papermaker and pasteboard maker who operated Hatfield Mill in Hertfordshire and a manufactory at Islington in North London, between 1803 and 1839. The Islington factory was relatively unusual in that it was powered by a horse-driven wheel. His papers and boards, which ranged in colour from a warm white to a deep buff, and in finish from a very high glaze to a very rugged ROUGH surface, were renowned and the rougher surfaces were particular favoured by De Wint among others. Very few were watermarked, in fact the only known dated Creswick mark reads CRESWICK / 1818. He often stamped his papers with blind embossed stamps in the corners, the most usual form being an oval bearing the words IMPROVED / DRAW^G PAPER / ROUGH / THOMAS CRESWICK.

34. There were many attempts at making a board that replicated the ivory tablets used by painters of miniatures and several recipes survive. The Society of Arts received "Mr Enslie's process for making Ivory Paper" in 1819, where it was noted down by Bryan Donkin. In his MS *Recipe Book*, c1820, we also find another recipe for a preparation for an 'Ivory' Board:

> Make a paste with fine flour or starch. When it is sufficiently boiled, add to it common turpentine in such quantity as shall be equal to about a twelfth part of thestarch or paste. Stir them well together and let them simmer over the fire for 5 or 6 minutes. Use the paste hot.

35. eg: The group of Palmer's watercolours based on Milton's *L'Allegro* (three subjects) and *Il Pensero* (five subjects), dating from 1868-70.

36. The normal way of exhibiting 'Exhibition' watercolours during this period was to mount them in a gold mount, right up to or overlapping the edge of the painted image. It is interesting that Palmer seems to be recommending a white mount: quite simply it allows the colours to stand on their own.

37. A. H. Palmer, *The Life and Letters of Samuel Palmer*, London 1892 (1972) Letter XLII to Miss Louisa Twining, November 22nd 1864. The 'whitey-brown' that Palmer refers to was originally a strong wrapping paper, made from a blend of white linen rags and old hemp rope, and which ranged in tone, depending on the maker and the batch, from a warm cream to a pale buff-grey.

38. A. H. Palmer, *op cit*, Letter LXXXIX to Mr P G Hamerton, November 1872.

39. A. H. Palmer, *op cit*, Letter LXXXVI to Mr Richard Seeley, November 6th 1872.

40. "James L Turnbull and J Turnbull, Makers of Playing Cards, Pasteboard, Paper Glossers and Pressers and Drawing Board Makers (including London Board)" in Bradshaw's *List of Papermills in England, Scotland and Northern Ireland*, Manchester, 1853.

41. Trade catalogues in the Winsor & Newton Archives.

42. Examples found include (shape of stamp indicated): TURNBULL / Arms of London / SUPERFINE / LONDON BOARD - vertical oval. BRISTOL / Crown / PAPER (for lightweights) - circle. BRISTOL / Crown / BOARD (for heavier weights) - circle. EXTRA SUPERFINE / Arms of London / LONDON BOARD - horizontal oval. EXTRA / SUPERFINE / Crown / DRAWING / BOARD - vertical oval.

43. RSA Archives, uncatalogued drawings entered for various prizes. Two contrasting examples, one made up only with Whatman paper and the other made up using no Whatman paper at all but both bearing the Turnbulls' blind-embossed stamp, TURNBULL / Arms of London / SUPERFINE / LONDON BOARD, on a vertical oval. The

first is a two-ply made up of two sheets, both watermarked J WHATMAN / 1827 and the second, a three-ply made up of three sheets each watermarked RUSE & TURNERS / 1839.

44. Letter to his son, the artist David Cox junior in 1845. N. Neil Solly, *Memoir of the Life of David Cox*, London 1873, p 131.
45. Solly, *op cit*, p 81. I have, as yet, not been able to discover which mill this would have been.
46. Grosvenor, Chater and Company, Wholesale Stationers, (founded by William Grosvenor c1690), occupied No 11 Cornhill, London, from 1765-1851 before moving to Cannon Street. Grosvenor Chater did not start making papers themselves until the 1850s when they acquired Abbey Mill, Flintshire (1854) and Glory Mill, Buckinghamshire (1859). They had previously had papers made for them by various makers including both Balston at Springfield Mill, Maidstone and the Hollingworths at Turkey Mill, Maidstone.
47. The only Dundee Mill I can find records of did not begin operation until 1850. It is possible that the story of a Dundee origin for this paper is incorrect. The Excise number 84B is also unusual, this is the only example I have seen with a letter in addition to the usual number.
48. The flecks come from the larger hemp fibres from old tarred ropes.
49. Trenchard Cox, *David Cox*, Phoenix House British Painters Series, 1947, p 82.
50. Birmingham City Art Gallery.
51. Advice given by Turner to Mary Lloyd, and recollected by her in "A Memoir of J.M.W. Turner, R.A. by Mary Lloyd", reproduced in *Turner Studies*, Vol 4 No 1, pages 22-23.

APPENDIX: WATERMARKS FOUND IN PAPERS USED BY WILLIAM BLAKE (1757-1827), JOHN CONSTABLE (1776-1837) AND JOHN SELL COTMAN (1782-1842).

Many other artists from this period could have been used for a listing such as this, but Blake, Constable and Cotman have such different ways of working, and such different aspirations, that a consideration of their usage is instructive. Comparisons of the lists show considerable overlap in the names of the paper-makers present. None of these three artists was as prolific or travelled as widely, either in Britain or abroad, as J M W Turner (1775-1851) so the range of papers available to them was necessarily more limited[1]. Three other artists working during this period, Thomas Girtin (1775-1802), David Cox (1783-1859) and Peter De Wint (1784-1849), could also have been chosen for this listing. Each made particularly idiosyncratic, and very interesting, choices in their papers but in the case of all three men, identification of the makers of their favourite papers is problematical.

Although some of the watermarks listed below appear to match, it does not mean that they are necessarily the same paper. For example, there are some forty different papers watermarked 1794 / J WHATMAN, more if we consider variations between batches of papers nominally of the same type[2]. The type of mould, furnish, finish, pressing, felts used, sizing, weight and size of sheet vary considerably. Each change in any one of these parts of the process imparts a subtle difference to the character and behaviour of the sheet in use. Those papers which have been identified as being the same are marked * in each list.

These three listings are based on my own examination of works by the three artists, supplemented by further information found in publishing records. Several authors have recorded information about watermarks found in papers used by different artists, but, as none are versed in paper historical research and as such information was often of only secondary importance to the real nature of their researches they have recorded their information in sometimes quite idiosyncratic ways. Although there are sometimes one or two misreadings of individual marks, these listings are of great use and the following appendix draws much on this earlier research.

Marks indicated thus ? are uncertain readings.
= signifies watermark and countermark present.
[   ] indicates missing part of mark, where known.

A: Blake's papermakers: Watermarks found in Papers used by William Blake (1757-1827)[3]

The makers and their marks

Charles Ball, Albury Park Mill, Surrey.
    C BALL
William Balston, Springfield Mill, Maidstone, Kent.

J WHATMAN / 1813 *	J WHATMAN / 1815
J WHATMAN / 1818	J WHATMAN / 1819
J WHATMAN / 1820	J WHATMAN / 1821
J WHATMAN / 1824	J WHATMAN / 1825
J WHATMAN / 1826	J WHATMAN / 1828
J WHATMAN / 1831 *	J WHATMAN / 1832
J WHATMAN / 1808 ?	

This last mark possibly originates from the Hollingworths at Turkey Mill, (see later). After the dissolution of the partnership between William Balston and the Hollingworth Brothers, both companies continued making in different mills. They both had the rights to use the WHATMAN name in their watermarks. It is not yet possible to ascribe with certainty the J WHATMAN watermarks with 1807, 1808, dates. After some confusion between the two companies the two companies the Hollingworths adopted the additional words TURKEY MILL or TURKEY MILLS in their marks.

William Balston & the Hollingworth Brothers, Turkey Mill, Maidstone, Kent.

1794 / J WHATMAN	J WHATMAN / 1804

Richard Barnard, Eyhorne Street Mill Kent.
    R BARNARD / 1827
Ann Blackwell, Nash Mills, Hertfordshire.
    A BLACKWELL / 1798
Charles Brenchley, Pratling Street Mill, Aylesford, Kent.
    C BRENCHLEY / 1804
William Bridges, Laverstoke Mill, Hampshire.
    W BRIDGES / 1794
William Bridges had been in partnership with Joseph Portal who had died in 1793. He then went into partnership with John Portal. This is perhaps a rare 'transitional' watermark, from the period between the two

partnerships. The most common forms of watermarks from the two partnerships are PORTAL & CO or PORTAL & BRIDGES.

John Buttanshaw, West Peckham Mill, Kent.
    BUTTANSHAW   BUTTANSHAW / 1799
    BUTTANSHAW / 1802

William, John & Thomas Curteis, Carshalton Paper Mill, Surrey.
    CURTEIS & SON

William Dickie, Bookbinder, Paper Merchant and Stationer, The Strand, London.
    W D[ ]E / 1803   W D[ ]IE / 1804
No certain information has yet been found as to which maker Dickie employed to produce papers with his own mark. At this period William Dickie was using papers made by the Balston Hollingworth Partnership at Turkey Mill, Edmeads and Pine at Ivy Mill, William Gater of South Stoneham, Hayes and Wise at Padsole Mill. An undated W DICKIE watermark has been found bound up in the same book as a very similar paper watermarked SMITH / 1796. Comparisons between the two papers suggest that they may well be from the same source[4].

Robert Edmeads & John Pine, Ivy Mill, Maidstone, Kent.
    E & P   EDMEADS & [PINE]
    EDMEADS & PINE / 1802   IVY MILL / 1806

William Elgar, Chafford Mill, Penshurst, Kent.
    Fleur-de-Lys / WE = W ELGAR / 1796

John & Edward Gater, Up Mill, South Stonham, Hampshire.
    GATER / 1805

John Green, Hayle Mill, Maidstone, Kent.
    J GREEN / 1819

F Hayes, Boughton Mill, Northamptonshire.
    F HAYES / 1798

John Hayes & John Wise, Padsole Mill, Maidstone, Kent.
    [HAYES] & WISE   HAYES & WISE / 1799

N Hendon.
    N HENDON / 1802
No papermaker of this name has as yet been identified. Hendon was possibly a stationer or paper merchant.

T & J Hollingworth, Turkey Mill, Maidstone, Kent.
    J WHATMAN / TURKEY MILL / 1821 *
    J WHATMAN / 1808 ?

Martha & Joseph Lay, St Mary Cray Mill, Kent.
    M & J LAY / 1816

William Lepard, Hamper Mill, Watford, Hertfordshire.
    [LE]PARD

John Pine & William Thomas, Basted Mill, Wrotham, Kent.
    BASTED MILL / 1820

Joseph Ruse, Richard & Thomas Turner, Upper Tovil Mill, Maidstone, Kent.
    RUSE & TURNERS   R & T
    RUSE & TURNERS / 1810   RUSE & TURNERS / 1812
    RUSE & TURNERS / 1815

Thomas Smith & Henry Allnutt, Ivy Mill, Maidstone, Kent.
    SMITH & ALLNUTT / 1815

Thomas Stains, Foots Cray Mill, Kent.
    T STAINS   T STAINS / 1813

John Taylor, Poll Mill, Maidstone, Kent.
    I TAYLOR   1794 / I TAYLOR

William Turner, Chafford Mill, Penshurst, Kent.
    W TURNER & SON

James Whatman the younger, Turkey Mill, Maidstone, Kent.
    J WHATMAN
    J WHATMAN = Fleur-de-lys / Strasbourg Bend

Henry Willmott, Shoreham Mill, Shoreham, Kent.
    H WILLMOTT / 1810

Watermarks from unidentified makers

AP   Circled F
HP   S[   ]
[ ]TH

B: Constable's papermakers: Watermarks found in Papers used by John Constable (1776-1837)[5]

The makers and their marks

William Balston, Springfield Mill, Maidstone, Kent.
    J WHATMAN / [ ]   J WHATMAN / 1810
    J WHATMAN / 1811   J WHATMAN / 1813 *
    [J WH]ATMAN / [18]18   J WHATMAN / 1829 *
    J WHA[ ] / 182[ ]   J WHATMAN / 1830 *
    [J WHA]TMAN / [18]31 * J WHATMAN / 1832
    J WHATMAN 1833 *

William Balston & the Hollingworth Brothers, Turkey Mill, Maidstone, Kent.
    1[79] / J WH[ATMAN]   1794 / J WHATMAN
    J WHATMAN / 1804

John Budgen?, Dartford Mill, Kent.
    J BUDGEN / 1820
    This is unlikely to be Budgen as he was a prisoner for debt in the Kings Bench Prison for most of 1820. He died in the October of that year. Whoever was working Dartford Mill was still using Budgen's watermarks.

John Dickinson, Nash Mills, Hertfordshire.
    JOHN DIC[   ] / 18[ ]

Robert Edmeads & John Pine, Ivy Mill, Maidstone, Kent.
    E & P   E & P / 1801
    E & P / 1802

John Gilling and Robert Allford, Cheddar Mill, Somerset.
    GILLING & / ALLFORD / 1829

John Hayes & John Wise, Padsole Mill, Maidstone, Kent.
    Posthorn / H&W

T&J Hollingworth, Turkey Mill, Maidstone, Kent.
    J WHATMAN / TURKEY MILLS / 1817
    TURKEY MILLS / J WHATMAN / 1819
    J WHATMAN / TURKEY MILL / 1821 *
    J WHATMAN / TURKEY MILLS / 1824

John Pine & William Thomas, Basted Mill, Wrotham, Kent.
    BASTED MILL / 1824

John Portal and William Bridges, Laverstoke Mill, Hampshire.
    PORTAL & CO / 1794

Joseph Ruse, Upper Tovil Mill, Maidstone, Kent.
    J RUSE / 1800

Robert Tassell, Lower Mill, East Malling, Kent.
    R TASSELL / 1833

James Whatman the younger, Turkey Mill, Maidstone, Kent.
    J WHATMAN

Thomas Wickwar, West Mills, Newbury, Berkshire.
    T WICKWAR / 1801

Watermarks from unidentified makers

B & M   Britannia
[ ]E   Crown / GR
Ornamented Cartouche
(Prob. James Whatman the younger)
Posthorn / M   W S[   ] / 17[ ]
1796   [ ]05

Constable also used several examples of "Bath Wove Post" writing paper, form unidentified makers, with its typical blind-embossed Crown / BATH stamp.

C: Cotman's papermakers: Watermarks found in Papers used by John Sell Cotman (1782-1842)[6]

The makers and their marks

William Balston, Springfield Mill, Maidstone, Kent.
    J WHATMAN / 1813 *   [J] WHATMAN / [1814]

J WHATM[ ] [J WHA]TMAN / [182]9 *
J WHATMAN / 1830 * J WHATMAN / 1831 *
J WHATMAN / 1833 *
William Balston & the Hollingworth Brothers, Turkey Mill, Maidstone, Kent.
 1794 / J WHATMAN
Robert Edmeads & John Pine, Ivy Mill, Maidstone, Kent.
 E&P / 1796
John Green, Hayle Mill, Maidstone, Kent.
 J GREEN / [ ]
T & J Hollingworth, Turkey Mill, Maidstone, Kent.
 J WHATM[AN] / TURKEY [MILL] / [ ]
John Larking, Upper Mill, East Malling, Kent.
 [J LA]RKING = Ornamented Fleur-de-lys / GR
Richard & Thomas Turner, Upper Tovil Mill, Maidstone, Kent.
 RUSE & TURNER / 1834
James Whatman the younger, Turkey Mill, Maidstone, Kent.
 JAMES W[HATMAN]
 The use of the Christian name, JAMES, rather than the more usual initial J, shows that this piece of paper has cut down from a much larger sheet, either *Double Elephant* (approximately 26 3/4 x 40 inches, 680 x 1019 mm) or *Antiquarian* (approximately 31 x 53 inches, 788 x 1347 mm). The Whatman watermarks found in these sheet sizes usually read in full: JAMES WHATMAN TURKEY MILL KENT and are generally dated.

Watermarks from unidentified makers

Crown [ ] [ ] WR4 monogram
Strasbourg Bend [1]800

Very few watermarks have been found in works by Cotman. Many of his works are laid down onto backings and have not yet been properly examined, and from those that have it is fairly obvious that, particularly on wove papers, he liked to work away from the watermark, which, in English papermaking practice at least, was generally placed along the edge of the sheet, was trimmed off after working.

Many of the buff laid wrapping papers favoured by Cotman earlier in his career were never watermarked and their origin is impossible to determine with any accuracy. Most of the buff woves he favoured later in his career can be identified, despite the lack of watermarks, as having been made by George Steart at Bally, Ellen and Steart's De Montalt Mill, near Bath, Somerset. The most usual form of the watermarks found coloured woves from De Montalt Mill is B, E, &S. + a date.

Examples have been found of Cotman working on unwatermarked sheets which have blind-embossed stamps:
 'London Board', from an unknown maker, stamped LONDON SUPERFINE BOARD.
 Buff wove drawing paper, from Thomas Creswick, stamped DRAWING PAPER / ROUGH / THOMAS CRESWICK in a circle.

Notes and References

1. for Turner's usage see Bower, *Turner's Papers: A study of the Manufacture, Selection and Usage of his Drawing Papers*, London 1990, where paper usage in the first half of Turner's career is examined.
2. eg; Known 1794 / J WHATMAN watermarked papers include the following, some of which are found in three different finishes, and all in different weights:
   Thin Post Writings, both laid and wove
   Thick Post Writings, both laid and wove
   Large Post Writings, both laid and wove
   Demy drawing, wove
   Demy printing, wove
   Foolscap Writings, both laid and wove
   Medium, writing
   Medium, drawing
   Royal writing, both laid and wove
   Royal drawing, wove
   Royal printing, wove
   Super Royal drawing, wove
   Imperial writing, wove
   Imperial drawing, both laid and wove
   Imperial Printing, both laid and wove
   Small Imperial drawing, wove
   Double Elephant, printing, wove
   Double Elephant, drawing, wove
3. Published sources for watermarks in papers used by William Blake: G. E. Bentley, jnr. *Blakes's Books*, Oxford 1977, pp71-3 provides a table of watermarks. His *Supplement to Blakes's Books*, 1995, provides an extra 12 examples of watermarks. The text volume of Martin Butlin's *The Paintings and drawings of William Blake*, 1981, p627 provides a further listing. Further information on some watermarks found in Blake's monochrome engravings can be found in *Essick's Printmaker*, 1980, pp71, 105, 160, and 257.
4. These two papers can be found together in a copy of *The British Itinerary* by Nathanial Coltman, used by J M W Turner as a sketchbook (*The Devonshire Coast No 1* sketchbook, Tate Gallery, Turner Bequest CXXIII). This book was published and bound by William Dickie and Turner's copy had been bound up with extra blank sheets interleaved amongst the printed text. Dickie had bound many of Turner's custom-made sketchbooks.
5. Published Sources for some watermarks found in papers used by John Constable: Graham Reynolds, *Catalogue of the Constable Collection in the Victoria and Albert Museum*, London, 1973. Jane McAusland, "Some of the Drawing Materials Available to Artists in the late Eighteenth and Early Nineteenth Centuries", in *Constable, a Master Draughtsman*, London, 1994. pp13-15.
6. Published Sources for watermarks found in papers used by *John Sell Cotman*: Miklós Rajnai and Marjories Allthorpe-Guyton, *John Sell Cotman, Drawings of Normandy in Norwich Castle Museum*, Norwich, 1975. Miklós Rajnai and Marjories Allthorpe-Guyton, *John Sell Cotman, Early drawings in Norwich Castle Museum*, Norwich, 1979. Andrew Moore, *John Sell Cotman, 1782-1842*, Norwich 1982.

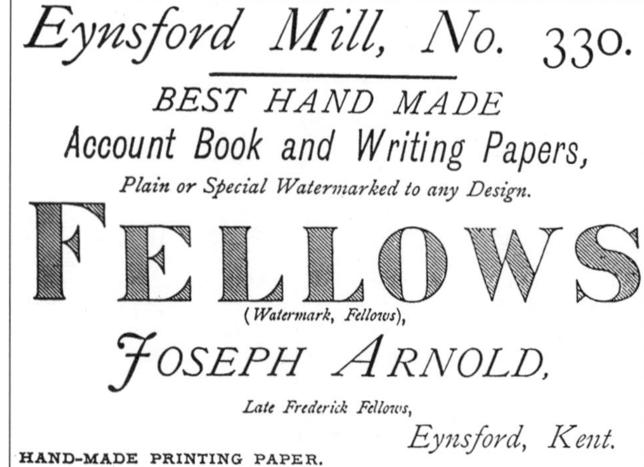

FIBRES AND PAPER

H.W. Kropholler, D.J. Ryder and C.P. Wilkins

Introduction

Paper grades are commonly marketed in seven categories namely newsprint, printings and writings, packaging papers, packaging boards, corrugating case material, tissues and speciality grades.

Table 1 was constructed after examining a cross-section of paper grades taken from a variety of the British Paper and Board Manufactures.[1] Consistent trends concerning the qualities and technical aspects of production, for the seven grades have emerged.

There is an apparent lack of importance attached to the types of fibre used to produce these papers. It is in house that pulps are carefully chosen by leading producers and so the high cost of pulp prevents the producers from disclosing information on preferred pulp sources. Nevertheless for many of the above grades a wide variety of pulp types can be used.

From the early 1940s to the mid 1970s an enormous amount of research was dedicated to establish how fibre characteristics influence paper properties. Normally a standard pulp evaluation procedure is done to establish the particular merits and qualities of a pulp and thus how it is likely to behave in paper. Indeed these techniques have been applied to assessing the potential of new pulp sources. However the problem with this approach is that it requires significant quantities of pulp.

Basic fibre properties had been thought to be capable of predicting sheet qualities. It is suggested in the literature that it is possible to relate certain basic fibre properties to sheet characteristics. It is evident that the choice of fibre property varied, and in some cases was contradictory, as to what property it was influencing. The conflict between researchers has undoubtedly arisen due to the fact that few if any of the basic fibre properties measured are truly independent. Figure 2 summarises which researchers used what basic fibre property to predict sheet behaviour. In some way basic fibre properties contribute to and influence the behaviour of the sheet, and must therefore be an important consideration when designing a product.

Fibres made from different materials and different processes do not have the same basic fibre properties. How strong the paper is, whether it will change with time and other properties will depend on the type of fibre used. In simple terms softwoods, hardwoods and

TABLE 1: QUALITIES AND ASPECTS OF CERTAIN PAPER GRADES		
Grade	**Needs**	**Technical Aspects**
Newsprint	Fast Cheap Opaque Printable	Mechanical and waste paper usually twin wire machine No additives usually
Printings and Writings	Opaque Smooth White Printable	Blended, bleached chemical pulps some mechanical pulps many additives, fillers, size etc.
Packaging Papers and Boards Corrugated	Cheap Strong Rigid	Waste and mechanical, some kraft softwood for strength. Often strength additives
Tissues	Fast Cheap Soft	Bleached recycled softwoods. Properties obtained from machine type
Speciality	Variable	Large contribution from machine type. Special additives. Fibre type often very important.

mechanical pulps can be separated into their contributing factors as follows:

> Softwood; higher strength (burst, tear, tensile) lower opacity and poorer formation.

> Hardwood; good formation and opacity, lower strength properties.

> Mechanical; very good bulk but subject to shade degradation, lower strength properties.

There are many factors which contribute to a paper grade. For example the paper machine can be taught to control the sort of paper produced. Good tissue paper cannot be produced on a newspaper machine, and so on. The total furnish is also very important since many of the additives used: fillers, binders, dyes, strength additives, are determined by the product requirement and in many cases actually make the paper grade itself. This can be quantified and one can see, in general, that the fibre only contributes some 20-25% to the paper type and quality.

Figure 1 is a pie chart representation of the contributing factors which go to make-up a paper grade.

A simple and quick way of characterising new sources of fibre would be to cook a small quantity (≤ 10 g) of the

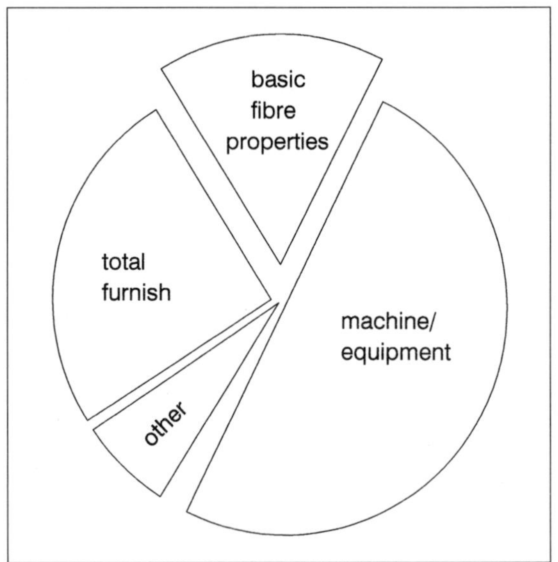

Figure 1: *Estimating the weighting of the factors which make-up a paper grade*

pulp source and then determine its basic fibre properties.

Below, results are presented for a variety of commercial pulp grades. The idea behind analysing already known grades of pulp is two-fold. These results and methods have been taken from Ryder.[21]

1) Establish whether the results from image analysis are sensible.

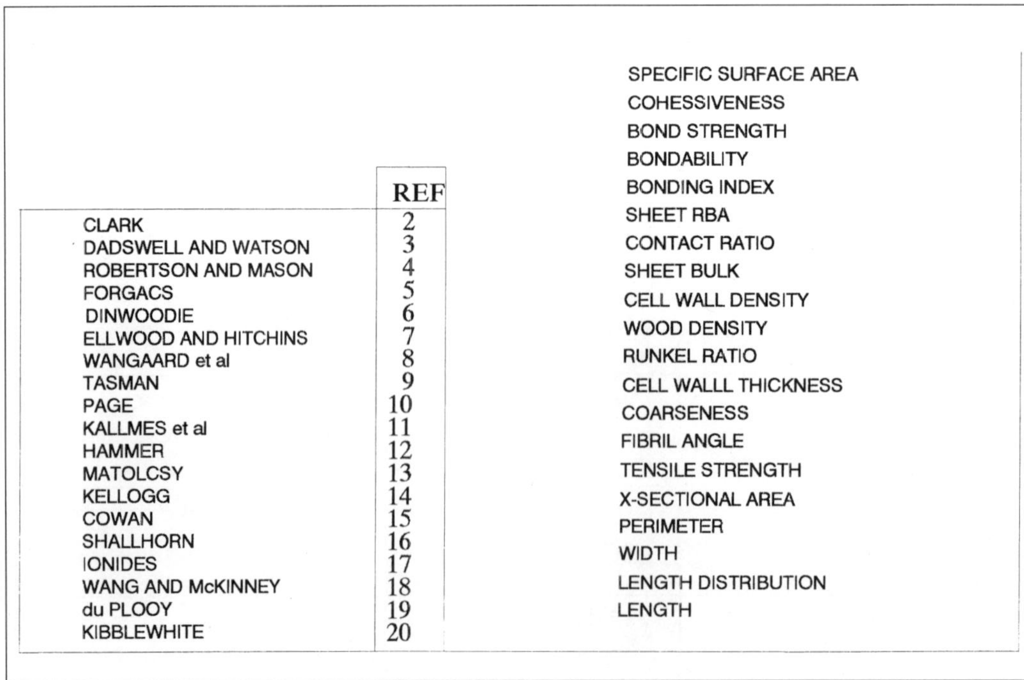

	REF
CLARK	2
DADSWELL AND WATSON	3
ROBERTSON AND MASON	4
FORGACS	5
DINWOODIE	6
ELLWOOD AND HITCHINS	7
WANGAARD et al	8
TASMAN	9
PAGE	10
KALLMES et al	11
HAMMER	12
MATOLCSY	13
KELLOGG	14
COWAN	15
SHALLHORN	16
IONIDES	17
WANG AND McKINNEY	18
du PLOOY	19
KIBBLEWHITE	20

Figure 2: *Summary of what basic fibre properties have been used to predict sheet characteristics*

2) The results for the commercial pulps can be used as a bench mark. Fibre and paper properties and the way they interact can be studied so that when a new fibre source with similar or unusual characteristics is come across it can be earmarked for further investigation.

Materials and Methods
A known weight of fibre is deposited onto a filter paper on the wire of a British standard sheet former. These fibres are then transferred directly to glass slides by couching in a handsheet press. The substance of this sheet or network will be 1-2 g.

The low grammage networks are then dried and stained to enhance optical visualization.

Fibre length and curl are determined using standard image analysis techniques, the proportion of area occupied by fibre is determined under both incident and transmitted illumination and the number of fibres per unit area counted.

From the above measurements it is possible to calculate the following basic fibre properties:

MEASURED BASIC FIBRE PROPERTIES
Fibre Length (mm)   500 measurements
Fibre Curl Index   ”   ”
Area Coarseness (gsm) weight per unit area
Decigrex (mg/100 m)   weight per unit length
Fibre Width (μ)   length/area

Fibre Thickness (μ)   area coarseness/cellulose density
Contact Ratio   incident area/transmitted area

British standard handsheets were made and tested (ISO) for a wide variety of pulp types.

MEASURED HANDSHEET PROPERTIES
Apparent Density (gcm^{-3})
Air Permeability (ml min^{-1})
Zerospan (Nmg^{-1})
Tensile Index (Nmg^{-1})
Burst Index (Nm^{-2}g)
Tear Index (mN.m^{-2}.g)

Results
Tables 2 and 3 show the basic fibre properties and the sheet physical characteristics, for a variety of fibre types.

Discussion
The length, width and thickness results (Table 3) can be used to describe fibre morphology, which in turn can be related to the sheet physical properties.

Fibre morphology not only affects the physical strength, it also affects air permeability and opacity. For example air permeability is greatest with the more rounded fibre morphologies and least with the flattened. An even greater reduction in air permeability occurs when there are multiple fibre types present e.g. flax and straw added to longer fibres. The small cells of Flax and Straw act like fines filling between the other fibres and thus produce a

TABLE 2: BASIC FIBRE PROPERTIES						
Fibre Type	Length (mm)	Width (μ)	Thickness (μ)	Contact Ratio	Area Coarseness (gsm)	Decigrex mg/100m
SOFTWOOD	3.0	35	3.4	0.45	5.1	9.9
HARDWOOD	0.9	49	2.5	0.28	3.7	18.5
SISAL	2.5	9	5.4	0.57	8.2	7.6
SUNN HEMP	1.5	24	3.6	0.75	5.4	12.8
ABBACA	3.5	19	4.6	0.51	7.0	13.7
ESPARTO	1.3	11.5	4.0	0.30	6.0	7.0
FLAX	1.0	20	2.7	0.65	4.2	8.4
STRAW	**0.9**	**38**	**2.1**	**0.80**	**3.2**	**12.1**
COTTON	2.0	8.5	2.7	0.27	4.2	9.5
SYNTHETIC	5.2	20	15.0	0.37	22.2	30.65
CTMP	2.0	21	6.8	0.40	10.2	20.9

TABLE 3: PHYSICAL PROPERTIES OF HANDSHEETS (ISO STANDARDS)						
Fibre Type	Apparent Density (gcm^{-3})	Air Perm. ml/min	Zero Span km	Tensile Index Nm/g	Tear Index mN m/g	Burst Index N/cm^2/g
SOFTWOOD	0.535	42	198	32	21	0.184
HARDWOOD	0.530	48	159	24	6.8	0.116
SISAL	0.437	72	158	28	22	0.179
SUNN HEMP	0.526	14	150	30	21	0.221
ABBACA	0.430	75	144	26	30	0.143
ESPARTO	0.370	76	148	19	3.9	0.058
FLAX	0.481	1.0	134	42	8.6	0.253
STRAW	**0.758**	**1.7**	**122**	**37**	**3.4**	**0.276**
COTTON	0.377	72	127	12	15	0.065

highly bonded pad of low air permeability. Incidently, the rate at which a pulp drains is also influenced in much the same way.

Opacity which is the light scattering ability of paper depends on the number of air fibre interfaces, the more interfaces the greater the opacity. Fibre morphology dictates the amount of air fibre interfaces. Fibres which can closely conform i.e. flattened or have lots of small cells tend to have a lower opacity because there are less air-fibre interfaces and more fibre-fibre interfaces. Conversely, pulps which have more rounded fibre types which cannot associate quite as closely tend to have better opacities. Moreover, those pulps which have a greater number of smaller less closely fitting fibres tend to have greater opacities because there are more fibre-air interfaces per unit area.

Figure 3: *Morphology of various papermaking fibres*

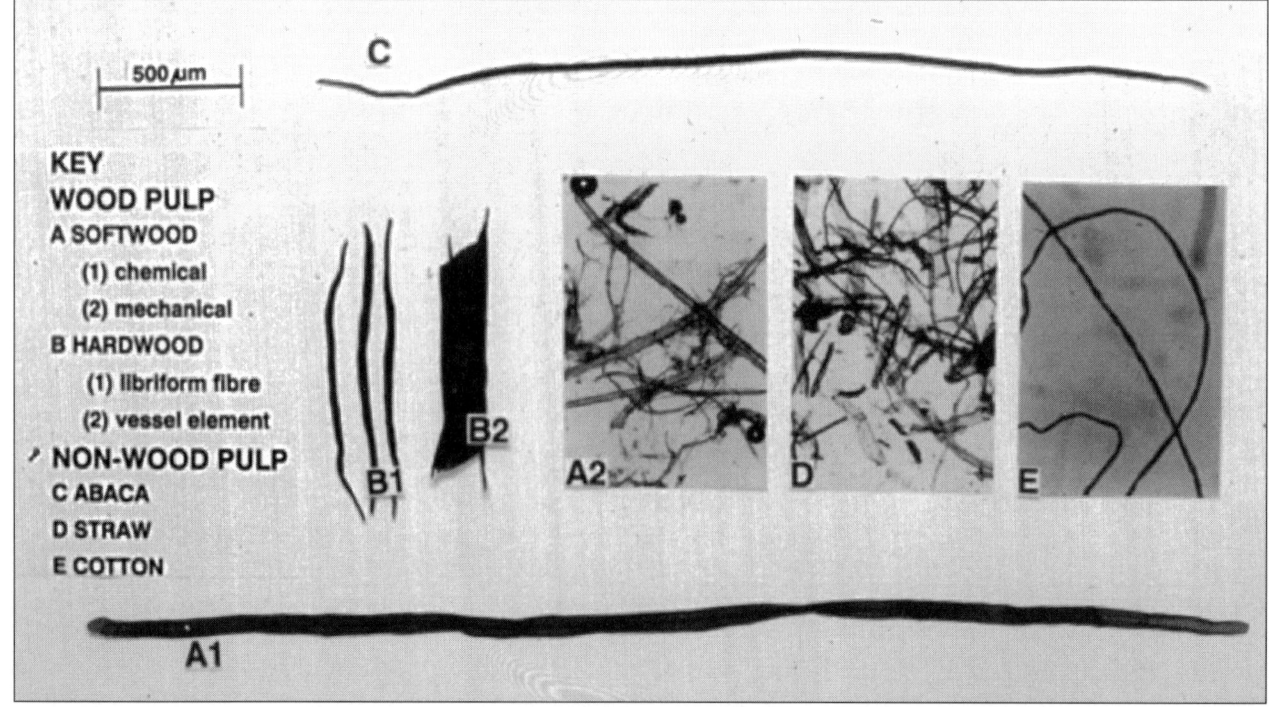

Kropholler et al: Fibres and Paper

FIBRE TYPE	AIR PERMEABILITY
FLATTENED MORPHOLOGY SOFTWOOD AND SUNN HEMP	LOWISH AIR PERMEABILITY 42.6 AND 14.0 ml/min
MORE ROUNDED MORPHOLOGY SISAL AND ESPARTO	HIGH AIR PERMEABILITY 72.4 AND 76.4 ml/min
MULTIPLE CELL TYPES FLAX AND STRAW	VERY LOW AIR PERMEABILITY 1.0 AND 1.7 ml/min

Fibre bonding is perhaps the most important fibre characteristic, without it paper would not really exist. Fibre morphology greatly influences the amount of fibre surface area available for bonding.

A flattened fibre morphology e.g. Softwood and Sunn Hemp, thickness 3.4 and 3.6 µm respectively, will provide more fibre surface for bonding and a stronger paper should result, as seen by the tensile index of 32 and 30 Nm/g.

Conversely a more rounded fibre morphology will bond less full and thus form a weaker sheet, e.g. Esparto and Sisal, thickness 5.4 and 4.4 µm and tensile index 19.2 and 28 Nm/g.

Obviously a longer wider fibre will have more area for bonding than will a short slim fibre and this must also be considered because it will also influence strength characteristics.

A more complex situation develops when a pulp source has more than one fibre type. Hardwoods generally have two main types of fibre: vessel elements and true fibres. The true fibres are characteristically thicker walled slimmer and longer than the vessel elements, their morphology is therefore more rounded and thus contribute less to the bonding than the vessel elements do. Flax and Straw have many cell types contributing to their respective pulp types. For the most part the flattened parenchyma and collenchyma act as bonding agents filling between the other cell types and increase the strength performance of these pulp types, thickness 2.7 and 2.1 µm and tensile index 41.7 and 37 Nm/g.

The chemical nature of the fibres have not been investigated, however the present techniques do permit some simple conclusions about fibre constituents, e.g. the contact ratio of Cotton versus Esparto. The chemistry of straw is extremely important in its potential role as a papermaking raw material. In addition to the work shown here the chemical composition of straw has an important role to play. The internode of the straw stalks contain some 3% silica. This material causes abrasive problems in the papermaking process. In chemical pulping various silicates may be formed. These silicates are related to cements. Straw tends to make a brittle paper.

Conclusions

Basic fibre properties influence sheet properties. However the amount of influence is dependent upon the whole papermaking process. The use of additives, refining and the type of machinery used to make the paper can often mask the influences fibres are exhibiting within the paper. Despite this certain grades of paper will still rely heavily on the initial qualities of the pulp.

For example, cotton is a highly desirable papermaking fibre and is used to produce high quality long life security papers. The fact that cotton is a very pure form of fibre enhances its durability and longevity in the paper product. It also has unique tactile qualities which are of importance to the security of the paper. Moreover, cotton still has to be mechanically treated and have a lot of chemical additives before it can be made into security papers.

The essential feature of natural fibres are represented by the high length to diameter ratio and the fact that natural fibres are essentially cellulose chains, make all fibre a potential renewable resource for paper grades.

In short the assessment of basic fibre properties for future pulp sources is very useful. The information provided can be used when deciding the performance requirements of the particular grades of paper. However availability of the raw material and the economics of production will probably dictate what papers are produced from any particular source.

Acknowledgements

This paper is based on a presentation of a paper "Do fibre types make paper grades" presented at the CELLULOSE '91 Conference held in New Orleans U.S.A. December 1991. The authors would like to thank Natalie Milowych for supplying information on the chemistry of straw.

References

1 Limes, C. and Booth, G., *Paper Industry Resource Pack*, published by Paper Publications Ltd. 1990.
2 Clark, J.d'A, *Tappi 56*, 7, 122-125 (1973).
3 Dadswell, H.E. and Watson, A.J., *Proc 2nd. Int. Fundamental Research Symp., Oxford*, 537-564, published by BPBIF, London, England (1962), Ed. F. Bolam.
4 Robertson, A.A. and Mason, M.G., *Proc. 2nd. Int. Fundamental Research Symp., Oxford*, 639-647, published by BPBIF, London, England (1962), Ed. F. Bolam.
5 Forgacs, O.M., *Pulp and Paper* Mag. of Canada, Convention Issue, T89-115 (1963).
6 Dinwoodie, J.M., *Tappi 49*, 2, 57-67 (1966).
7 Barefoot, A.C., Hitchins, R.G. Ellwood, E.L. and Wilson, E.H., *Tech. Bulletin 202*, N. Carolina Agricultural Experimental Station, N.C. State Uni. Releigh, N. Carolina (1970).
8 Wangaard, F.F., Kellogg, R.M. and Brinkley, A.W. *Tappi 49*, 6, 263-277 (1966).
9 Tasman, J.E., *Pulp and Paper* Mag. of Canada, T553-569 (1966).
10 Page, D., *Tappi 62*, 9, 99-102 (1970).
11 Kallmes, O.J. Bernier, G. and Perez, M., *Paper Tech. and Ind.* 18, 8, 22-27 (1977).
12 Hammer, R.J., *Paper Tech.* 15,5, T153-160 (1974).
13 Matolocsy, G.A., *Tappi 58*, 4, 136-141 (1975).
14 Kellogg, R.M. and Thykeson, E., *Tappi 58*, 4, 131-135 (1975).
15 Cowan, W. and Cowdrey, K.J.K., *Tappi 51*, 2, 90-93 (1974).
16 Shallhorn, R.M., and Jones, A.Y. and Karnis, A., 62nd Annual Meeting, *Proc. Tech. Section* A493, C.P.P.A., Montreal, Canada (1976).
17 Ionides, G.N. and Mitchell, J.G., *Paperi Ja Puu 60*, 4a, 233-238 (1978).
18 Wang, P.H. and McKinney, M.D., *Tappi 60*, 7, 140-143 (1977).
19 du Plooy, A.B.J., *Appita 33*, 4, 257-264 (1980).
20 Kibblewhite, R.P. and Barley, P.G. *Pro. Phy. Confer.* (1987).
21 Ryder, D.J. *PhD Thesis* UMIST 1992.

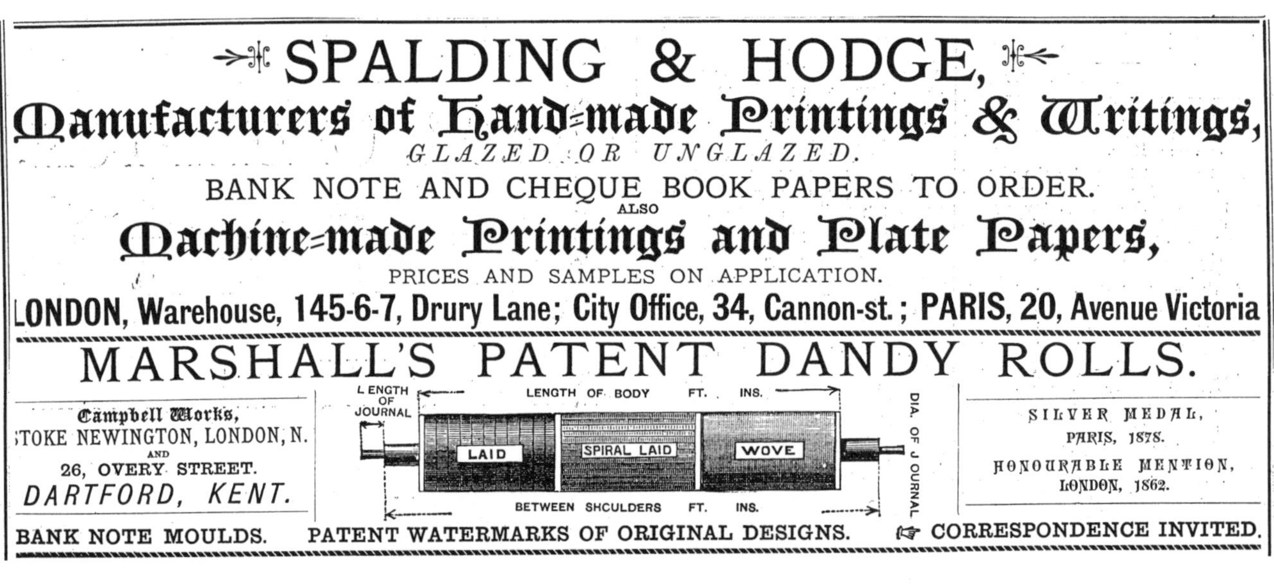

JAMES WATT AND HIS COPYING MACHINE

Richard Hills

After his move to Birmingham in 1775, James Watt produced a much more economical pumping engine with the successful application of his separate condenser. Sales began to increase from around 1777 onwards and Watt found that, not only did he have to be away from Birmingham much more frequently but that the volume of correspondence also increased. For example, he had to spend long periods in Cornwall supervising the erection and maintenance of engines and it is doubtful whether he had any secretarial assistance while he was there. Also records of correspondence with customers needed to be kept in Birmingham so that reference could be made to past letters and agreements in case of any dispute. At this period, either everything had to be written by hand or set up in type and printed on a printing press. While copies produced on the printing press would be identical, this form of reproduction was economical only where a reasonable number of copies was needed. When only one other copy was wanted, a duplicate was written out by hand with all the possibility of mistakes by the person making the copy.

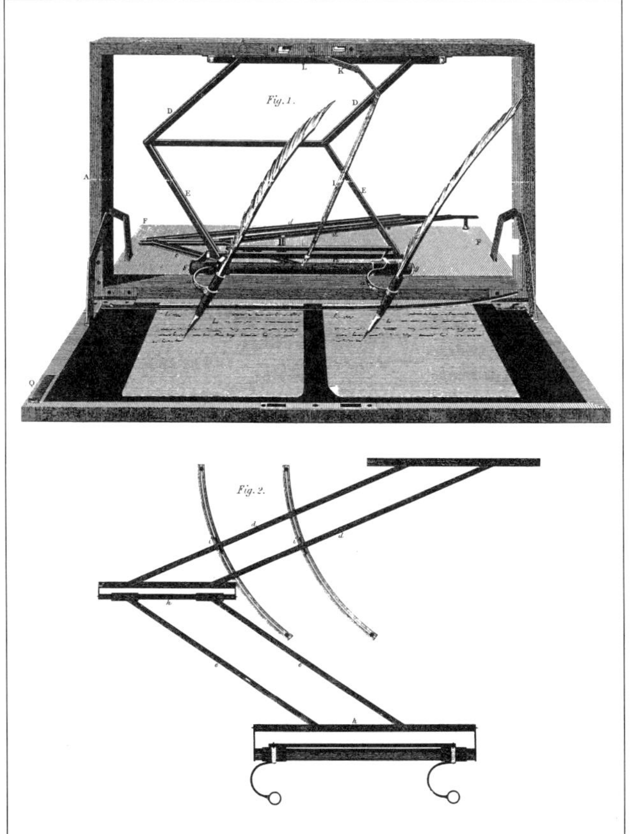

Figure 1: *Hawkin's Patent Polygraph, as illustrated in Rees'*Cyclopedia *of 1819.*

Watt's interest in a copying device is said to have been aroused by Dr. Erasmus Darwin who proposed a sort of double pen which he called a "bigrapher".[1] Rees's *Cyclopedia* has a description and illustration of a similar sort of device, Hawkin's patent polygraph.[2] A frame held a couple of quill pens, one of which was guided by the writer on one sheet of paper and the other followed on a second sheet. It was claimed that up to five pens could be used and the instruments "answer the purpose very well". But a similar device was "found to take considerably more than twice the time to produce its two copies that the common pen took to produce one".[3] Watt must have wondered whether he could invent something better.

Some of the test trials carried out by Watt on small pieces of paper have been preserved in his Garret Workshop in the Science Museum in London. An examination of these showed that here we have evidence for an advance in the history of papermaking and that we are able to learn a great deal about the introduction of a new use for paper, a very rare occurrence. This report is a brief summary and may need revising in the light of further research.

Watt must have been working on his ideas during 1779 for on June 28 he wrote presumably to Matthew Boulton:

> I send you enclosed some of Mr. Nobody's draughts with authentic copies of them; you will observe that that which appears full black will not preserve even the colour it has but will grow browner by keeping. The paler kind are nearly if not quite fully as white as the paper was before the operation and will stay so.
>
> I will be much obliged to you to procure me a quire or two of the most evenly & whitest *unsized* cambric paper, and also specimens of the cheaper kinds with their wholesale prices. If you apply at Curtis's [you] may perhaps be better served as we have dealt with him before. It is absolutely necessary that there by no size in the paper which [you] may know by touching it with a wet finger.
>
> The copies will continue to grow blacker than they were before copying and as far as I can judge not in the least defaced.[4]

This letter suggests that Watt was having difficulty both in finding the type of paper he wanted and also with the colour or keeping qualities of the ink. In fact he was to spend a great deal of time over many months trying to perfect his process to make it into a viable marketable operation. He must have been pleased with his progress with his experiments for on 24 July, 1779, he wrote to Joseph Black in Edinburgh:

> I have lately discovered a method of copying [writing] instantaneously, provided it has been written [the same day] or within 24 hours. I send you a specimen [and will] impart the secret if it will be of any use to you. [It enables] me to copy all my business letters.[5]

From this letter it would seem that Watt had achieved a reasonably good system but he continued to experiment for a further year, even after he had taken out a Patent Specification for his invention. To see what he was doing, let us turn to his Patent Specification for some of the details. The Patent Specification is dated 14 February 1780 and was signed by Watt on 22 May that year and enrolled on 31 May 1780. I give these dates because the trial test pieces of paper show that Watt was continuing with his experiments on both ink and paper right through this period and into November 1780. I have not been able to discover whether any alterations were made in the Patent Specification between submission that February and its sealing in May.

Watt stated,

> Let the letter, or other writing that is intended to be copied, be written with the ink hereinafter described, or with any other writing ink fit for the purpose. Take a piece or pieces of thin paper which contain no size, or glue, or gummy mucilaginous matter, or which at least does not contain so much size or other matter as would make it fit for being written upon. Cut this paper or papers to the size and shape of the writing of which a copy is wanted. Moisten or wet the said thin paper with water, or other liquid, by means of a sponge or brush, or by dipping, or otherwise. Having moistened or wet the thin paper, lay it between two cloths, or other substances capable of absorbing the superfluous moisture from the thin paper. When it has been slightly pressed between such thick spongy paper, or other substances, by the hand or otherwise, lay the said thin paper, so moistened and pressed, upon or under the side of the writing which is to be copied, and in such manner that the one side of the said moistened paper shall be in contact all over the side of the said writing so intended to be copied; and that to the other side of the said moistened thin paper there shall be applied a piece of clean writing paper or cloth, or other smooth uniform substance. Lay the said writing intended to be copied

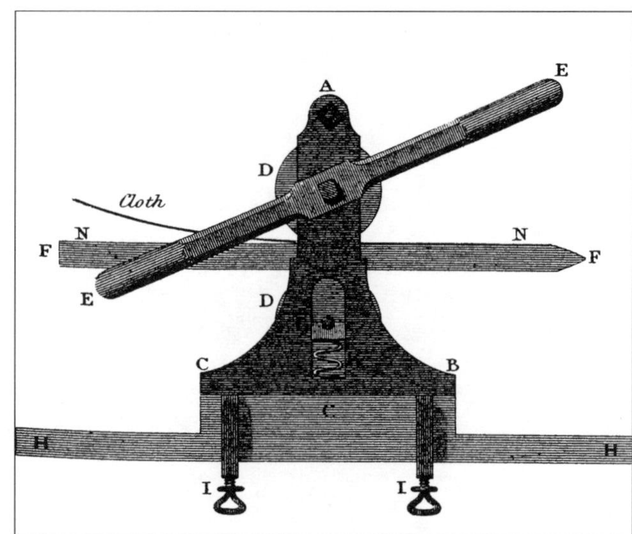

Figure 2: *End view of James Watt's Rolling Press, from Rees'* Cyclopedia *of 1819. The text that accompanies this illustration reads: ABC is one of the ends of an iron or wooden frame, which serves to connect the two rollers. D, D, are two wooden or metalline rollers, turned extremely exact, or truly cylindrical, and which are mounted on iron axles, firmly fixed in them. E E is a doubled-ended lever, by means of which the roller, on whose axle it is applied, may be forcibly turned round. F F represents the board of the rolling-press, on which the writings to be copied are to be laid. N N is a piece of cloth, or other elastic pliable substance, placed next the roller, and above the writings to be copied; and the board, G, is a strong plank of wood, or plate of metal, serving to connect the two end-pieces of the frame at the bottom. H H represents the edge of a common table, to which the press may be fastened by the iron screw-cramps I,I. K is a slit, of which there is one in each end-piece of the frame; these slits are filled with elastic steel, or other metalline springs, or with some other elastic substances which serve to press the two rollers forcibly together. L is a brass bolster, supported upon the springs, and serving to support the end of the axis of the under roller.*

with the thin moistened paper intended to receive the copy, placed respectively as above directed, upon the board of a common rolling-press, or of that of which a description and drawing are hereunder written and drawn, and pass them once, or oftener, through the rolls of the said press, in the same

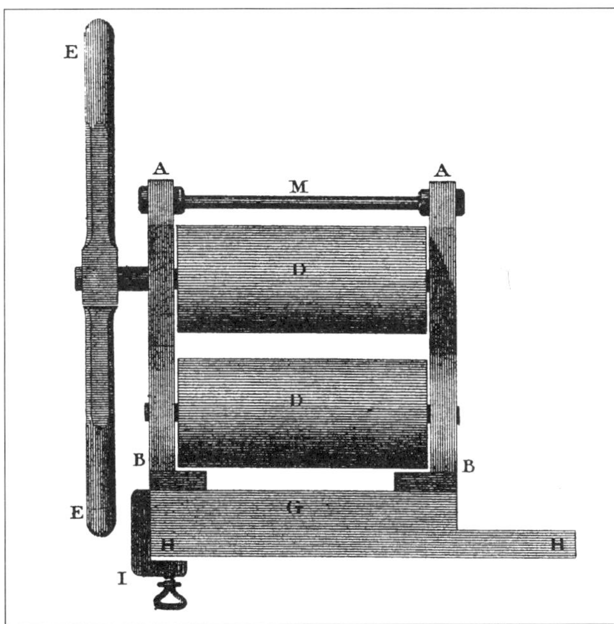

Figure 3: *Side view of James Watt's Rolling Press, from Rees'* Cyclopedia *of 1819. The text that accompanies this illustration reads: A B, A B, are the two end-pieces of the frame. D, D are the two rollers. E is the double-ended lever. G is a strong plank, or plate of metal, which forms the bottom of the frame. H H is the table on which the press stands. I is one of the iron cramps which fasten the press to the table; and M is a bar of iron which connects the upper part of the frame.*

manner as is used in printing by copper plates; or... subject them to any other pressure sufficient for the purpose, by means of which pressure, in whatever manner applied, part of the ink of the writing intended to be copied shall pass from the said writing into, upon, and through the said thin moistened paper, so that a copy of the said writing, more or less faint according to the quality of the ink and paper employed, shall appear impressed on both sides of the said moistened paper, viz. upon one of the sides in the natural or proper order and direction of the lines as they are in the original writing, and on the other side in the reverse order and direction.[6]

What Watt had done, in effect, was to moisten the ink again on the original letter so that it soaked right through the thin wet second sheet of paper and could be read in the normal way on the upper surface of that second sheet.

Watt described in his patent specification a preparation to be applied to the tissue paper to help enhance the copy and also how to make the ink. I am no chemist so am unable to assess the chemical constituents of Watt's mixtures to see how well they would work. There are two further problems. The first is that notes on the chemicals used on the pieces of test paper are written either in symbols for which I have found no key or in code such as *Soho* or *Newhall* for which again there is no key. The second problem is that Watt wrote to Black on 29 March 1780:

> You [seem] to apprehend that the Copying Scheme may be made a bad use of, that I shall take care to prevent by publishing ink which cannot be copied and paper which being wrote upon by any ink is also incapable of yielding a copy.[7]

Therefore we may have to suspect the formulae published by Watt because he may have tried deliberately to deceive people so they could not discover his secrets.

The problems of finding an ink which would retain its colour after the copying process took Watt a long time to achieve. In his Garret Workshop is a packet with two powders. The instructions say that the larger parcel had to be boiled for half an hour or longer and then, when it had cooled to blood heat, the contents of the second packet could be added. While this concoction could be used immediately, it was better left to stand for a couple of days during which time the colour would develop. So the preparation of the ink was quite laborious and even the chemist, Joseph Black had trouble with it. Watt's letter of apology to him tells us about some of the difficulties in its use.

> I am sorry the ink you got with the [copying] machine did not prove good. It was good when first made but from some cause not well understood yett by us, the Ink of which that was a part has deposited great part of its colour and become very thin. That subject is attended with more difficulties than you can be aware of - The following qualities are required - 1st that the Ink be capable of being made up in the form of a powder readily soluble in water - 2dly that it be fitt for use in an hour or two after it is made - 3dly that it be thin and flow readily from the pen - 4thly that it do not grow thick or mould by keeping; nor yett deposit much sediment - 5thly that it be capable of giving an impression without diffusion, soon after it is wrote with - 6thly that it shall be capable of yielding an impression at the end of 24 hours after it is wrote with and if it would do it at a longer date would still be more valuable, lastly that the Copy as well as the original keep it colour...[8]

From this we can see that the ink had to be capable of being used in the usual way on normal writing paper. I have not done a thorough survey of the paper Watt used but I can say that it was generally good quality. A little after 1780, he was using Whatman laid and wove paper. This, and presumably other good quality writing paper, would have been well sized with a gelatine size. Therefore I presume that the ink would not penetrate far into the actual paper under the size. In Watt's correspondence that I have read, there is no sign of the ink diffusing into the paper like blotting paper nor of it penetrating through to the reverse side. Presumably the ink needed to be fairly thick to leave a sufficient deposit from which some could be taken to make the copy.

Watt wanted to be able to take a copy as long as a day after the letter had been written so presumably the ink had to remain slightly moist but not so moist that it would smudge. The addition of sugar or gum was one way of achieving this and Watt commented on one of his test samples, "Soho + Sugar runs too much out of the pen - new extract + Gum writes very well, Soho + logwood + @ writes thinner than it did". When the damp sheet of copy paper was placed over the writing and pressed onto the original sheet, the ink had to transfer without diffusion and penetrate through this second sheet of paper so that the writing could be read on the further surface the correct way round. The paper for this had to be thin and even. This damp paper was sandwiched between the dry sheet to be copied and another dry sheet of either thin "Calimanco", presumably a thin cotton cloth, or ordinary paper or blotting paper. The purpose of this may have been to draw the moisture out of the copying paper and help the ink to penetrate better. There is a note on one test sample, "To try the effects of Calimanco in place of the Blotting papers as they suck too much".

We have seen that Watt asked Boulton to find some unsized paper, but thin tissue paper when damped must have become weak and very difficult to handle so that it is possible that a lightly sized paper as mentioned in the Patent was more generally used. A well beaten pulp would have given a stronger sheet when damped. In the Garret Workshop there is a pair of small wooden boxes or trays and lids. Both lids are lined with foil and the underside of one box. The bottom of this box appears to be made from plaster of paris. This could have been soaked with water, the copy paper placed in the box and the lid put on to enable the paper to become saturated. The tray of the second box was lined with foil and the damped paper could have been stored here until wanted. In fact Black had problems with the paper and wrote to Watt, "After bragging so much of my Ink I have spoiled my letter by takeing a Copy with too wet a paper".[9] That the paper was almost as crucial as the ink for a successful copy is shown by a letter from a William Duncan also preserved in the Garret Workshop:

> When I bought one of your Copying Presses, soon after you gave them to the public, I received with it, half a Ream of *very fine*

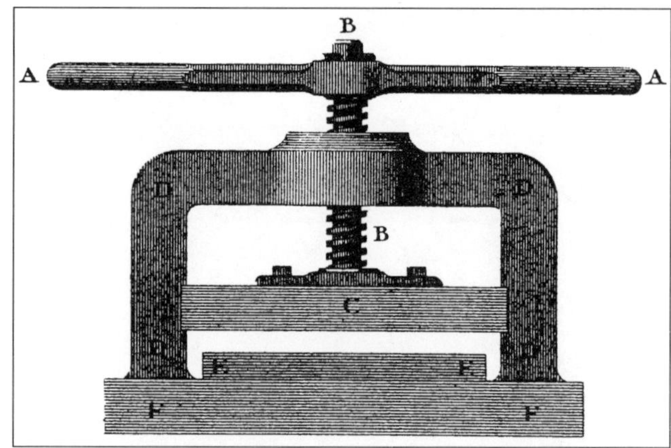

Figure 4: *James Watt's Screw Press, from Rees'* Cyclopedia *of 1819. The text that accompanies this illustration reads: Screw-press, which may be used, instead of the rolling press, in taking off impressions from writings. A A is a double-ended lever. B B the screw. C a block of wood, or metal, which the screw acts upon, and which is attached to it. D D the frame of the press, made of iron or wood. E E is a moveable board, on which the writing to be copied is to be laid, with a cloth over it. F F the bottom or sole of the press, made of wood or metal. Be it remembered, that these presses are made of a different sizes, according to the sizes or largenesses of the writing intended to be copied. Those now referred to are drawn from one sufficiently large to take an impression from a folio page of writing or post paper, and are drawn to a scale of one inch and a half for each foot, or one-eighth of their natural size.*

Copying paper from Mr. Woodmason, since which I have not been able to get any but what is *very thick course and uneven.* I shall therefore esteem it a favour, if you will let me know the reason of this alteration in the paper and how it is to be remedied as I hear many complaints of the same kind from numbers of my friends to whom I have recommended your useful invention.[10]

Mr. Woodmason was a stationer in Leadenhall Street who was one of the main suppliers of ink and the actual copying machine.

On some of the pieces of paper he used for his tests, Watt wrote the name of the manufacturer as well as notes about the chemistry. To me it is surprising to find so many papermakers making such lightweight papers and I have wondered what were the other uses to which it was put at this period besides covering printed plates in books.

Another use was lightweight in two senses. Ephemeral literature such as popular ballads which were not normally kept for long were printed on lightweight paper so that the pedlars could carry bundles of them around the country more easily. Perhaps of more significance was the printed transfer paper used by potters to decorate their wares. Watt, with his involvement in the pottery business, must have been aware of the use of this thin paper.

It would seem that all these tests were carried out during 1780, mostly during March and April after Watt had made the submission for his patent but some start that January and some continue into November. Most of the early tests were carried out on laid paper. However, on such lightweight paper, the chain lines particularly make very thin and weak places in the sheets and writing or letters on these are often difficult to read. Some laid samples have chain lines spaced very closely at only half inch intervals. On one is written "Kidderminster" and on another dated July 18, "Hurcott Paper". This must be Hurcott Mill and it is interesting that heavier paper with the same chain line spacing, which is unusual, was used by Baskerville to print an octavo edition of Virgil's Aeneid in 1766. Did this paper come from the same mill and so did Baskerville use Hurcott paper? Watt also used Whatman laid paper with a chain line spacing around ¾ ins. apart. One bundle of this Whatman paper had been used for tests in March and another in probably September. On one sheet Watt wrote, "Despond not, Whatman, perhaps it may do at last". The tests on paper from other manufacturers suggests it did not and there is no sign of any Whatman wove, at least none is so labelled, because we find Watt using wove paper from other manufacturers. Wove paper would have a more even density and thickness without the thin places caused by watermarks or chain lines.

In April Watt was using a laid (chain lines 11/16 ins.) "Taylors white thin paper" and also Taylors Opake which was wove. On April 18, there are samples of Taylor's paper in both laid and wove and more of Taylor's wove on August 26. Might this be James Taylor who had inherited Poll Mill, Kent from his father Clement who had died in 1776? In 1785 the mill was bought by James Whatman and James remained as tenant for a couple of years.[11] Returning to Watt, in April 1780 he was trying "Biggs's new paper no preparation" and also "Biggs's new paper weak preparation", both of which were laid but on July 17, "Biggs's paper" which was wove. On a piece of wove, undated, Watt wrote, "This is written to try Mr. Biggs new thin paper of which I have just received a ream from Mr. Woodman it seems very good as far as can be judged from appearances, Ink No. 14, the pen bad". The Biggs family was running Iping Mill, Sussex, at this time, a white paper mill with six vats, while a John Biggs is recorded by Shorter at Scots Bridge Mill, Hertfordshire in 1796.[12] Later samples are more generally on wove paper and some of the copies of letters in the Wilson Papers in Cornwall County Record Office have been made on wove paper watermarked along one edge J WATT & CO PATENT COPYING I noticed this on a letter dated 15 December 1792 but there could be much earlier examples because I was not looking for watermarks at the time.[13]

The J. Watt & Co. refers to the partnership which was established between Watt, Matthew Boulton and James Keir, a chemist who must have helped with the experiments on inks. The shares were; James Watt one half, Boulton and Keir one quarter each but the last bore all the expenses of the patent.[14] Black was quick to actively canvass the idea in Edinburgh and soon obtained orders from various people including the Duke of Buccleugh but they had to wait until the patent was sealed. The price of the licence was to be fixed at £6 (£5. 15. 6 in a later prospectus) and Watt must have hoped to obtain many orders because he printed a prospectus in which it was stated that the subscription lists would be closed after one thousand orders had been received until these were completed.

In May 1780, everything must have been ready for the launch of the copying machine and Watt sent Black thirty copies of this prospectus which he asked him to give to any of his friends who might buy a machine and "also to gett 2 or 3 of them stuck up in some Capital Booksellers Shops - We do not give any premium for taking in the profits as the paper will prove a sufficient indemnification".[15] Boulton, with his usual enthusiasm took a machine to London and showed it to members of both Houses of Parliament, bankers, and to frequenters of coffee-houses. He succeeded in arousing great interest at Westminster, so much so that "The Speaker... was often obliged to send his proper officer to fetch away from me the members to vote and sometimes to make a House".[16] The machine was even demonstrated to the King.

The actual machine consisted of either a screw or roller press, the roller press being made with rollers up to 16 inches long costing half a guinea an inch of the length of the rollers. Watt envisaged some form of bulk production because he wrote, "The Machine alone would cost more money [than the £6] if we did not make a great number together for we mean to make them very handsome and good".[17] Black and the Duke of Buccleugh were among the first to receive machines and that October Black reported that he had helped set up the one for the Duke of Dalkeith. The Duke was away but he "Shewed it to the Duchess and I hear they have since tried it with success and are very pleased with it".[18]

The prospectus claimed that "By means of this Invention, the practice of which is exceedingly easy, any person may take an exact copy of a letter or other sheet of paper written with common ink in a minute or less". It was a perfect copy with no omissions or additions as might happen by some one writing a duplicate. The apparatus was small and could be made into a "small, handsome

Figure 5: *Copying sample sent out with Watt's Prospectus for his Copying Machine: "Time Labour & Money are saved, Dispatch & Accuracy are attained, and Secrecy is preserved by this newly-invented Art of Copying Letters and other Writings."*

piece of furniture in a gentleman's study". Drawings of an elegant table on which to place the machine could be supplied by Watt. On the prospectus was stuck a sample of the copying paper with copied onto it the words, "Time Labour & Money are saved, Dispatch & Accuracy are attained, and Secrecy is preserved by this newly-invented Art of Copying Letters and other Writings".[19] Watt also used this method to copy engineering drawings. In this case, the thin copy paper was too weak to handle in such large sizes so the copy was transferred onto ordinary drawing paper. The ink did not penetrate to the further side so these copies remained backwards and were marked "Reverse". They were made without lettering or figures which had to be added later.

Possibly through the difficulty of preparing the ink as well as handling the moist paper, the copying machine was not an immediate success for Watt wrote in October 1780 to Black, "That business meets with Tolerable success though not so great as we had reason to expect".[20] By the end of the first year, 150 machines had been sold and orders were still coming in.[21] In another letter that November, Watt wrote again to Black "We have since made some discoveries in the nature of Ink since I last wrote to you on that subject and hope to do well with that article yet...".[22] Watt may have continued experimenting for there is another recipe for ink in a letter to Keir dated 15 January 1782.[23] However slow its acceptance at first, the copying machine based on Watt's experiments and models was to become one of the indispensable pieces of office equipment throughout the nineteenth century and well into the twentieth with its new use of the thin tissue-like copying paper. When Watt was replying to one of his correspondents in 1809, he wrote:

> It is gratifying that you find the copying-machine useful to you. It has been so much so

to me for the last twenty-six years that it has been worth all the trouble I had with it, had it been attended with no other profit.[24]

Author's Note

Recently more documents have been found in the James Watt Papers at Birmingham which give more information about the problems Watt had in procuring a suitable paper for use in his copying machine. These give details of different paper makers and show that several were producing wove papers in 1780. These documents will form the basis of a further paper.

References

1 Muirhead, J.P., *The Origin and Progress of the Mechanical Inventions of James Watt*, Murray, London, 1854, Vol. 3, p.27, see note.
2 Rees, A., *The Cyclopaedia, or Universal Dictionary of Arts, Sciences and Literature*, Longmans, London, 1819, Article on "Copying of Letters".
3 Muirhead, *op. cit.*, p.27.
4 Dickinson, H.W., *James Watt, Craftsman & Engineer*, Cambridge University Press, 1935, p.116.
5 Robinson, E. & McKie, D., *Partners in Science, Letters of James Watt and Joseph Black*, Constable, London, 1970, Letter 52, Watt to Black, 24 July 1779.
6 Muirhead, *op. cit.*, pp. 30-33.
7 Robinson & McKie, *op. cit.*, Letter 65, Watt to Magellan, 20 March, 1780.
8 *ibid*, Letter 73, Watt to Black, 25 October, 1780.
9 *ibid*, Letter 72, Black to Watt, 18 October, 1780.
10 Letter in Garret Workshop, Science Museum, London, From Will. Duncan, Philpot Lane, London, 20 July, 1782.
11 Shorter, A.H., *Paper Mills and Paper Makers in England, 1495-1800*, Paper Publications Society, Hilversum,

Hills: James Watt and his Copying Machine

1957, pp.188-9.

12   *ibid*, pp. 178 & 241.

13   *Wilson Papers*, Cornwall County Record Office, Truro, Vol.5, 15 December 1792.

14   Roll, Sir E., *An Early Experiment in Industrial Organisation, being a History of the Firm of Boulton & Watt, 1775-1805*, Cass & Co., London, 1968, p.130.

15   Robinson & McKie, *op. cit.*, Letter 68, Black to Watt, 17 May 1780.

16   Roll, *op. cit.*, p.131.

17   *ibid*, Letter 61, Watt to Magellan, 9 March 1780.

18   *ibid*, Letter 72, Black to Watt, 18 October, 1780.

19   Prospectus for the Copying Machine in the Garret Workshop.

20   Robinson & McKie, *op cit*, Letter 7, Watt to Black, 9 November, 1780.

21   Roll, *op. cit.*, p.132.

22   Robinson & McKie, *op cit*, Letter 75, Watt to Black, 9 November, 1780.

23   Dickinson, *op. cit.*, p.117.

24   Muirhead, *op. cit.*, p.28, note.

Appendix 1 : Transcript of the Advertisement for James Watt's Copying Machine

Proposals for Receiving Subscriptions for an Apparatus *by which* LETTERS *or other* WRITINGS *may be copied at once, and for the* LICENCE *of using the said Apparatus, an exclusive Privilege by his Majesty's Letters Patent having been granted to the Inventor for the sole use of his Inventions*

1. By means of this Invention, the practice of which is exceeding easy, any person may take an exact copy of a letter or other sheet of paper, written with common ink, in a minute or less.

2. From the nature of the method employed, the copy thus taken must be a perfect resemblance of the original writing: it is therefore not liable to faults of those copies that are transcribed, in which words are often from negligence, omitted, added, or altered; and hence it is much more valuable, not only as it is perfectly like the original, but also as it carries with it a testimony of its authenticity.

3. The apparatus will take up but little room, and may be fixed upon a desk or table in a compting house or upon a separate mahogany stand, so as to make a small, handsome piece of furniture in a gentleman's study. Proper instructions and drawings will be given to subscribers, so that any cabinet-maker or carpenter may fix it in any of these manners which may be required.

4. The necessity of keeping copies of letters of mercantile, and all other business is sufficiently known, and the conveniency and satisfaction of preserving copies of letters written on other subjects will be readily admitted. The *utility* therefore of a method by which writings may be copies exactly and almost instantaneously, must strike every person. To the *Merchant, Tradesman, and Lawyer*, this [   ]pply the place of a clerk, [   ] copying not only *letters* b[   ]s, *Bills of Parcels* and various *other* writings, and even *drawing*[   ]*ntleman* will hereby have an opportunity of preserving his [   ]nt letters with little trouble; Gentlemen who compose for [   ] by this means ob-tain a duplicate of their works before [   ]m to the Printer; Ambassadors, or other persons employe[   ]*airs* may thus re-tain duplicates of their most confidential[   ]out risk of disco-very by employing transcribers. In sh[   ] to whom *Time, Labour*, and *Expence*, are valuable and w[   ]n to write upon subjects in any degree interesting, will [   ]fit and pleasure from being possessed of this Invention.

5. Some persons have suggested that improper uses may be made of this art. Without entering into the question at present, it is sufficient to say, that no such practices can impose upon a person who makes use of this invention. Neither will any of *his* writings be liable to such practices, as will be more fully explained in the instructions that will be delivered to the subscribers along with the apparatus.

6. Subscriptions are taken at Mr. *James Woodmason's*, Stationery, *Leaden-hall-street*, Messrs. *Powney* and *Gust's*, Stationers, in *Parliament-street*, and at Mr. *Blamire's*, Corner of *Northumberland-street, Strand*, London.

Each subscriber is to pay £5. 15s. 6d. for which sum he is to receive an apparatus and a licence for using the Invention.

7. As soon as a thousand subscribers are received, the subscription is to be closed till the above are completed; and care shall be taken that each subscriber shall be furnished in order, according to the date of his subscription. Therefore those gentlemen who are desirous of being soon in possession of this Invention, will please to send their names, that they may be included in the first subscription.

In order to shew a specimen of this art, the following sentence is written in the usual way, and the copy taken from it is annexed below.

"Time, Labour & Money are saved, Dispatch & Accuracy are attained, and Secrecy is preserved by this newly-invented Art of Copying Letters and other Writings".

Appendix 2 : Watt's Instructions for using his copying ink

Patent Copying Powder, prepared and sold wholesale James Watt and Co. Birmingham

The quantity of powder contained in this paper is sufficient to make half a pint of ink; it consists of two parts put up separately. Put the larger parcel into a glass or stone pot, or into a ten ounce wide mouthed phial and pour on it half a pint of boiling hot water, taking care not to pour it on too suddenly lest it break the vessel. Stir the mixture and set the pot containing it into a vessel of boiling water, which keep boiling for half an hour or longer, or set the pot with the mixture for the ink so near the fire that it may boil for half an hour. When the brown powder appears to be wholly dissolved, remove it from the fire, and when it is about blood warm, add the smaller parcel of powder. Stir the whole well together and as soon as the powder is dissolved the ink will be fit for use: but it ought to stand in the open vessel in which it is made forty-eight hours, during which time it will deposit a sediment from which the thin ink should be poured off and kept in a phial [?well] corked.

The ink will be very pale and brownish when first made, but will grow black though it be used in its pale state. When [it is] newly made the impression taken from the writing by the Copying machine will be more liable to be diffused than after the ink has been kept some days in the ink pot; but [ ] writing made with it when new admits of [   ]

copied at a longer date after it is ?written [ ] afterwards and therefore those who do not write very much divide the [ ] parcels of powder into two parts and ?use only half the quantity of ink at a time.

The addition of a tea-spoonfull of good ?French Brandy to the half pint of ink helps to prevent it from moulding.

The ink made with this powder makes more [ ] writing and more *durable impressions* by means of the Copying Machine than any other and is possessed of the other qualities of the best inks.

Price 9d. per paper.

Sold by James Woodmason, Stationers, Leadenhall Street, London.

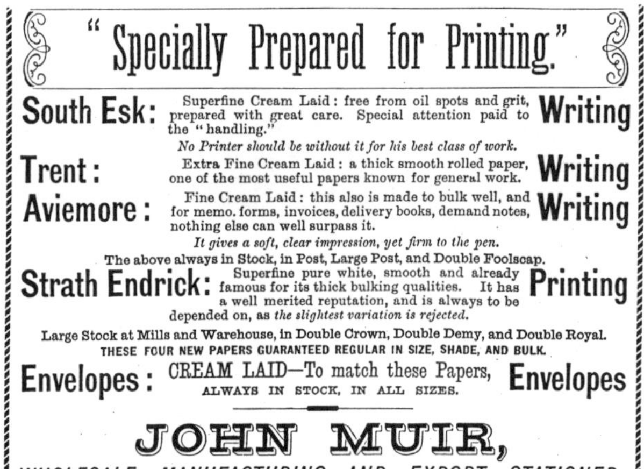

OBSERVATIONS FROM AN ART CONSERVATOR ABOUT THE USE OF STRAW IN PAPER AND BOARDS

Penny Jenkins

I have approached straw from two directions. One from the front, in the paper artefact itself which may contain processed and bleached straw, and one from the back, basic strawboard, onto which works of art are often stuck down.

Strawboards

The underlying blight from boards made from untreated cereal straws are lignin and other impurities which can cause staining and eventual degradation of the picture. So much of our work is stabilising pictures which have gone dark brown or have developed foxing spots because of exposure to light, damp and acidity while stuck down onto either a strawboard or mechanical woodpulp backing. Since boards became available in the late 19th century framers and mount makers have used incredible amounts of cheap board to stick down pictures to make them flat. According to one source[1] in the 1920s strawboards from Spicers Ltd. merchants were less than half the price of their woodpulp boards. They were sold in standard bundles weighing 56 lb. The heavier weight board, the less in the bundle. A large imperial (32" x 22") 8 oz board had 112 to the bundle. A 2½ lb board had 22 to the bundle etc. (Strawboard did not have to conform to the standardisation of paper and boards created in 1925). One dreads to think how many bundles of board were used by the framing trade over the last 100 years and the resultant damage will keep the conservation profession busy for at least another 100 years.

Figure 1: *Amateur watercolour by H. Stevens, from the 1930s? (exact date unknown), before treatment. The paper had an 'oatmealy' texture from a mixed fibre furnish with a high percentage of straw. It is stained and foxed in a manner to suggest the causes of degradation are inherent in the paper.*

Figure 2: *The same watercolour after treatment, including washing under alkaline conditions.*

The sorts of boards made from unbleached straw and waste pulps had the minimum decomposition process. After steeping in lime to break down the fibre, they were just boiled and beaten. Special board machines deposited the waste pulp onto a revolving cylinder at the end of the machine which received the web on a continuous basis. When the desired thickness was obtained, the wet sheet was slit and removed from the cylinder and taken to be pressed and rolled.[2] When removing these boards from the backs of pictures, you can observe the formation process. Some separate down into irregular layers and you end up with a hugh pile of sodden straw pulp remnants. They can be much more difficult to control than boards made on a multicylinder machine which can be separated and removed in coherent layers. Both these treatments can be painstakingly slow and difficult. Deconstructing strawboards is a very unglamorous but important skill in our profession! Another observation on strawboards is that they sometimes froth when wetted. This may be due to re-activating the insoluble soaps formed when lime was used as the initial alkaline decomposer. (Lime forms insoluble soaps with the greasy elements of the straw plant matter).[3] These same by-products are a hazard for paper making and machinery.

Cheap strawboards often have large lumps of recognisable straw left in them. Often these are the nodes at the leaf joints which take more time to break down.[4] These lumps can distort fine papers or become areas of surface abrasion when sheets get stacked one on top of another.

Straw in Paper
Straw in paper can be difficult to identify with certainty in some cases. Obvious shives of plant material may be tiny splinters of wood, weeds or sweepings from the floor.

A David Cox watercolour c. 1856, painted on a coarse wrapping purchased from Grosvenor Chater Merchants,[5] has a very lively fibre furnish which he has exploited by turning incidental tar specks into birds in flight. There is no doubt that Cox had chosen this paper for its warm buff tonality to provide contrast and a unique background texture. Naturally artists throughout history would use any sort of paper available and usually chosen on fairly arbitrary grounds based on price, colour and workability. Many artists chose warm toned straw papers to paint chalk and gouache studies and these media have a natural brilliance against the dark background.

The Cox had a fairly typical behaviour during treatment. The picture was firmly stuck onto two different acidic woodpulp supports and was stained and blotchy looking. When the final backing layer was released from the verso the watercolour sheet was very absorbent, especially in the areas with large straw flecks, suggesting that localised acidity had broken down whatever sizing was there originally. It responded well to a brief washing under very controlled conditions to reduce soluble staining and acidity. From experience I think degradation products from straw are more easily soluble in water than from woodpulp sources, and papers with a straw content generally respond well to minimum treatment. Sizing material appears to break down in papers made entirely

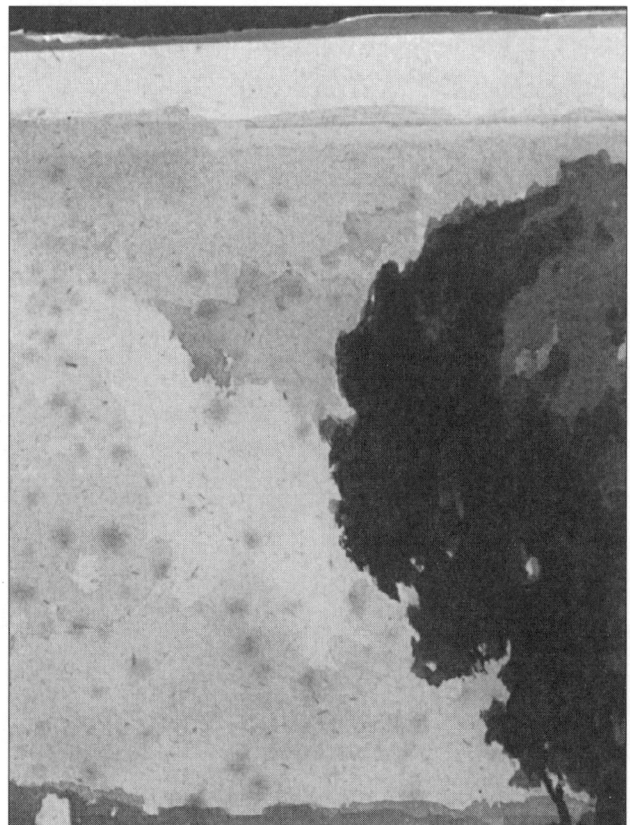

Figure 3: *Detail of watercolour before treatment*

from straw. The consequent absorbency and softness of straw papers can be hazardous. I have seen 19th century chalk drawings on all-straw laid paper which, if they were to get damp, could disappear completely, as the acidity released could dissolve the natural white chalk pigment made from Calcium Carbonate. How much sizing of straw paper was done is open to speculation. John Dickinson writing to Charles Longman in 1852, quoted in *The Endless Web* describes from a second hand observation that Mr. Joynson's paper which contained a large mixture of straw apparently took up size very well.[6] One conservation treatment if appropriate, is to re-size paper, usually by brush sizing with a warm gelatine solution. I would generally re-size straw papers after any wet treatment and they do noticeably gain a feel and handling strength.

Straw is probably far more common in papers used by artists than generally assumed despite its expensive process costs. Processed, bleached straw papers are more obviously in circulation in the 1930s to 1940s when there were wartime shortages of suitable paper making materials. Most of the papers have composite fibres with straw added to give bulk and a crispness to the paper. Many still look and handle well and their texture and colour give a distinctive background tonality to much early 20th century British art. From a conservation point of view, many works of art from this period are in fine condition if they have been kept in a good environment. However papers which have been subjected to light and dampness or have been stuck down onto acidic supports can look very degraded.

Straw papers tend to develop a distinctive decay pattern of large soft blotchy foxing spots. I am not sure if these are catalysed by the presence of straw or other impurities in the paper. On the positive side, these types of stains can respond well to simple conservation wet treatments, especially if under alkaline conditions. The fact that straw contains less lignin than woodpulp[7] must contribute to this phenomenon. If it were not for the appalling pollution implications in the manufacture of straw containing papers, they would certainly be preferable to some of the woodpulp papers that conservators have to deal with now and in the future.

References

1 Dawe, E.A. *Paper and its Uses, A treatise for printers stationers and others.* Crosby Lockwood and Son, London 1929. Volume 2. Sample description for *Boards.*

2 Dawe, E.A. *Paper and its Uses, A treatise for printers stationers and others.* Crosby Lockwood and Son, London 1929. Volume 1, p.31.

3 Grant, J., *A Laboratory Handbook of Pulp and Paper Manufacture*, Edward Arnold & Co., 1944, p.17.

4 Hills, R.L., *Papermaking in Britain 1488-1988*, The Athlone Press, London, 1988, p.136.

5 See Bower, Peter. "Information Loss & Image Distortion resulting from the handling, storage and treatment of sketchbooks and drawings" in *The Institute of Paper Conservation Conference Papers : Manchester 1992*, p.65, for more detailed account of Cox's purchase of this particular coarse paper which he called Scotch wrapping.

6 Evans, J. *The Endless Web, John Dickinson & Co. Ltd. 1804-1950*, Jonathan Cape, London, 1955, p.110.

7 Grant J., *A Laboratory Handbook of Pulp and Paper Manufacture*, Edward Arnold & Co., 1944, p.30. According to Table 2, cereal straws on average contain about 17% lignin compared to woods which contain about 27% lignin.

(Incidentally printed on paper which appears to contain flecks of straw).

W RAITT'S PHOTOGRAPHIC RECORD OF KASHMIRI PAPERMAKING IN 1917

Barry Watson

Papermaking is believed to have been introduced to Kashmir, situated in the north west Himalayas, from Samarkand during the fifteenth century AD. Some 500 years later papermaking methods were little changed. In 1917 the Indian Government asked the industrial chemist William Raitt M.I. Chem. E., F.C.S., to visit Kashmir and report on the possibility of updating the craft-based industry. Raitt was familiar with modern papermaking methods introduced from Europe and operating in other parts of the sub-continent, and he pioneered the use of bamboo as a source of paper-making fibre[1].

At this time, machine-made papers were already becoming available in Kashmir. Raitt's subsequent report concluded that there were no improvements to be made and that the ancient methods in use would soon pass into history. During his visit to Kashmir, Raitt took photographs of working mills, and in 1930, after retiring to the UK, he gave a lecture in London to a group of British papermakers[2]. At his lecture Raitt used photographs he had taken in 1917 and also demonstrated sheet forming with a Kashmiri flexible hand-mould that he had acquired. The photographs were probably taken in the *Arach* and *Nowshera* areas where papermills had operated for centuries. This paper is based on the 25 photographs reproduced in the published version of Raitt's text with an additional commentary.

The American paper historian, Dard Hunter, was also researching the ancient paper industry of India during the 1930s. His investigations culminated in a publication which included a detailed description of Kashmiri papermaking which confirmed and elaborated on Raitt's report[3]. This book is not readily available, but the same author's *Papermaking* gives a useful summary[4].

1. *Scene near a Pulp Mill*[5]. Hidden away in the north western Himalayas.

2. *A Pulp Mill.* Plentiful supplies of fibre in the form of old clothing (mostly cotton fibre) plus good water supplies are both available.

3. *A Pulp Mill.* A closer view of the pulp mill showing the waterwheel and the wooden troughs for controlling the water flow to the beaters and the vat.

4. *Another view of the pulp mill.* Showing the wheel.

5. *The Beater (left).* The rags are roughly sorted, torn into strips, and pounded (nearly dry) in this hammerhead beater. Where the rag is especially dirty, lime or wood ash soda may be added to assist cleaning.

7. *Beater Mortar (above).* A hollowed out stone used as a beater mortar. Notice a second mortar-like hole worn out by long use. Unlike the barred hollander, this equipment primarily fibrillates rather than shortens the fibres. Surprisingly strong paper can be produced from apparently poor raw material.

6. *The Beater. Man-powered, Ahmedabad.* Man power was used where water power was not available.

8. *Pulp Washer or Potcher.* Washing, or 'potching' using two-man power. The two men stand in the stream carrying the pulp held in a long piece of cotton fabric while the swift flow of water in the stream washes the pulp.

9. *Pulp Washing.* The potchermen use their hands to assist the washing process, in which dirt, colour, and some fibres, are removed.

11. *Pressing after Washing.* Further de-watering occurs as the half-stuff is squeezed in the fabric.

10. *Pressing after Washing.* After washing the two-man press effectively de-waters and consolidates the pulp (half-stuff).

12. *Pulp Drying and Bleaching.* The consolidated cakes of pulp are laid out to dry for several days.

13. *Pulp Drying and Bleaching.* Because of the altitude, the air has a high ozone content, and this, combined with intense sunlight, produces a marked bleaching effect.

14. *A Papermill.* As with earlier European handmade papermills, the papermaker and his family both live and work in the same building.

15. *A Papermill. The Water Supply.* Controlling the water supply to the vat. Stock is produced in large jars (see figures 9, 10 and 11). Concentrated rag pulp is put into a jar, with water, and an operator (an 'agitator') climbs inside using hands and feet to produce the mixed stock.

16. *Mould and Frame.* The deckle is made from cedar wood and has triangular cross-section ribs to ensure good drainage. The flexible mould cover is made from stems of a local grass sewn together using horses' tail hairs.

17. *The Vatman at Work*. A 'shake' is not normally used. Two wet sheets may be formed, one on top of the other, to form a duplex paper.

18. *The Vatman at Work*. The mould and deckle have been removed from the vat to drain.

19. *The Vatman at Work.* A newly-formed sheet on the mould, waiting to be couched.

20. *Couching.* The flexible mould cover has been turned over, with the wet sheet still adhering to it.

21. *Couching.* Couching in progress. The mould is carefully lifted leaving the wet sheet behind. A 'post' of 70-80 sheets is produced.

22. *Couching.* Here a second wet sheet is being couched onto the first sheet. This produces a heavier, stronger product.

23. *The Couch Press*. The five-man press. The 'post' will be left overnight weighted with heavy rocks to maintain pressure.

24. *Drying (Waterleaf)*. The 'post' is separated into wads of 6–8 sheets. Each wad is pressed against a smooth plastered south-facing wall. Surface smoothness tends to vary, but by drying several sheets together drying is more even and the likelihood of the paper being blown away is reduced.

25. *Sizing and Knotting*. On the left, the size-man, using a woollen mitt, applies a starch 'jelly'. The paper is then re-dried over rope lines, where it is examined for defects such as creases, knots, etc. Some sheets will be smoothed using a polished agate stone, held in a wooden handle, as can be seen on the right. Packing into bales and transporting the bales to the customer or bazaar, by horse or mule, completes the operations.

References

1. Raitt, W, *The Digestion of Grasses and Bamboo for Papermaking*, Crosby & Lockwood, 1931.
2. Raitt, W, *The Childhood of Papermaking as illustrated by Kashmiri methods of the Present Day*, Technical Section Proceedings, British Paper & Board Makers Association, Volume 11, 1930.
3. Hunter, Dard, *Papermaking by Hand in India*, New York, 1939.
4. Hunter, Dard, *Papermaking: The History & Technique of an Ancient Craft*, New York, 1947.
5. The picture titles are Raitt's own titles from his 1930 lecture.

Contributors

Peter Bower

Papermaker and Historian, currently continuing his researches into the Papers used by J.M.W. Turner, at the Tate Gallery. The first fruits of this research were exhibited and published in 1990: *Turner's Papers 1787-1820, A Study of the Manufacture, Selection & Use of His Drawing Papers*. Also involved in consultancy work for various galleries and institutions and the legal profession, involving the dating and attribution of documents and works of art on paper. Current researches include the investigation of a recently discovered hoard of 19th century banknote forgeries, correspondence, equipment, print trials and photographic material. Member of the Committees of the National Paper Museum Trust and The British Association of Paper Historians, editor of *The Quarterly*, The Journal of the British Paper Historians and member of the Board of the British Society of Questioned Document Examiners.

James Brander

Present job: Head of Exploratory Research at Arjo Wiggins Research & Development Limited. Time at Arjo Wiggins (WT) R. & D.: 13 years. Previously: Built printing machines with Cobden Chadwick. Education etc.: PhD in seismology - several years post-doc. Hobbies: Cycling, mountains, photography and woodwork.

Alan Crocker and Anne Phillips

Alan Crocker is Professor of Physics at the University of Surrey. He is a founder member of The British Association of Paper Historians and the present Chairman of the Association. He has published two limited edition books, *Paper Mills of the Tillingbourne* and *The Diaries of James Simmons, Paper Maker of Haslemere, 1831-1868*, several articles on the paper mills of Surrey and over 100 scientific papers. His other interests in industrial history and archaeology include gunpowder mills and water turbines. With his wife he is joint editor of the Newsletter of the Wind and Watermill Section of the Society for the Protection of Ancient Buildings.

Anne Phillips works at the National Physical Laboratory at Teddington, West London. She is the great-great-great-grand-daughter of Charles Ball III, papermaker of Surrey and Normandy, and has carried out extensive research into her family history in both England and France.

Phil Crockett

Phil Crockett is a graduate in mechanical engineering from Southampton University. He started in the paper industry when he emigrated to Canada in 1963 and joined Abitibi Power & Paper. After working in Toronto and at a Newsprint mill in North Ontario for 4 years, he returned to the U.K. and joined Wiggins Teape's Project Engineering Division.

In the last 25 years he has engineered projects in most of Wiggins Teape's mills for producing a wide range of fine, special and industrial papers, as well as plant for synthetic papers made with foam, and a production line for reconstituted tobacco.

He is currently working at Devon Valley Mill, and in his spare time has an interest in the history and development of paper-making, particularly the engineering aspects.

Dr. Richard Hills

Founding Curator of the Manchester Museum of Science and Technology, which incorporated the National Paper Museum. Past President of the International Association of Paper Historians and Past Chairman of the British Association of Paper Historians. Occasional lecturer at the University of Manchester Institute of Science and Technology. Author of various books on the history of Engineering and Technology: these include *Machines, Mills and Uncountable Costly Necessities; a short history of the drainage of the Fens, Power in the Industrial Revolution, Beyer, Peacock, Locomotive Builders to the World, Paper Making in the British Isles 1488-1988*, and *Power from Steam: a history of the Stationary Steam Engine*, To be published: *Power from Wind: A history of Wind Power*.

Penny Jenkins

Penny Jenkins is a freelance paper conservator working in London. She trained at Camberwell School of Art & Crafts and received a diploma in the conservation of prints and drawings in 1979. After working for a private collector and dealer she set up her own studio in 1981, and specialises in mostly western works of art on paper from the 16th Century to the present day.

An interest in some obscure printing methods, led her to write and research on 18th century 'Chine' papers and their European equivalents in the 19th Century, for making 'India' proof prints. She has also written about vellum and parchment papers for The Journal of the Institute of Paper Conservation.

Hein Kropholler

Read Chemical Engineering at the University of Cape Town. Worked in the paper industry in South Africa and Canada. Lecturer, Reader and Professor of Chemical Engineering at the Universities of Cape Town and Natal (South Africa) and Delaware (U.S.A.). Head of department of Paper Science UMIST from 1975 to 1990. Currently Professor of Paper Science at UMIST. Published some hundred papers on control, systems modelling and simulation, paper physics and applied microscopy.

Frances Wakeman

Frances Wakeman is a partner in the Plough Press which was started by her late husband, Geoffrey Wakeman. The Plough Press produces hand printed books in small numbers on the history of papermaking and other book arts. Through these activities an interest in the history and manufacture of paper led to a study of the Oxfordshire paper mills. This brief account of some of the mills is based on a more detailed book which is due to be published later.

Barry G. Watson

Barry Watson joined the paper industry over 50 years ago as a laboratory boy at John Dickinson & Co Ltd's Croxley Mill, Hertfordshire, later working in various production departments. In 1952 he moved to Albert E. Reed & Co Ltd (now Reed International) where he was responsible for the development and running of a wide range of pulp and papermaking courses in the company training school. These were attended by both company personnel as well as staff from outside client mills and organisations. Latterly he worked as a consultant for PIRA International and lectured to technical training courses both in the UK and overseas. A past examiner for the City & Guilds of London Institute and an honorary retired member of the British Paper Industry Technical Association. Co-editor and author of the 1979 textbook *Paper and Board Manufacture*. Has written several articles for *The Papermaker* (Hercules Powder Co, Wilmington, Delaware, USA) and the BAPH *Quarterly*.

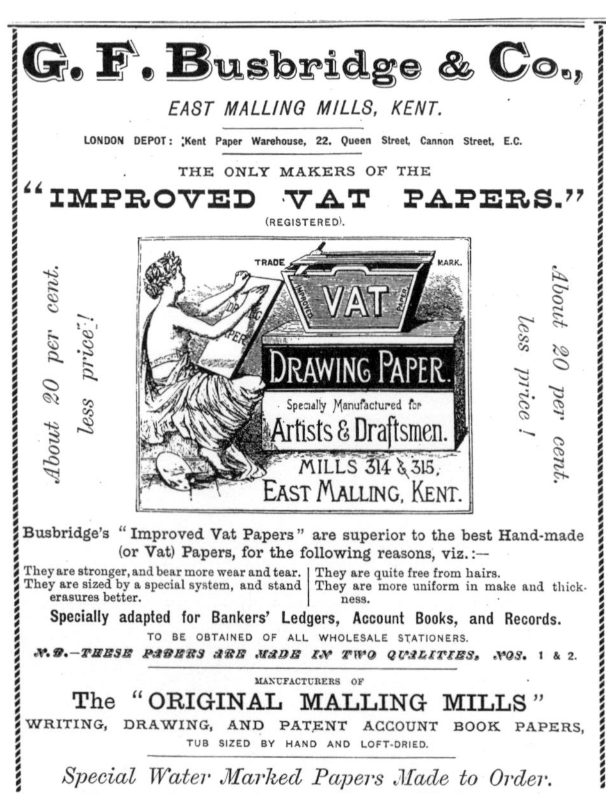

INDEX

The names of individual **Papermakers** and **Paper Mills** and descriptions of **Watermarks** can be found under those headings in this index. Entries covering other aspects of those subjects are listed after the alphabetical listing of the names. The **Papermakers** listing contains the names of makers, mill owners, companies, individual workers in the mills, and family members. Those **Watermarks** which are illustrated are marked thus *.